Aircraft
Accident
Analysis

Aircraft Accident Analysis

Final Reports

James M. Walters

Robert L. Sumwalt III

McGraw-Hill

New York   San Francisco   Washington, D.C.   Auckland   Bogotá
Caracas   Lisbon   London   Madrid   Mexico City   Milan
Montreal   New Delhi   San Juan   Singapore
Sydney   Tokyo   Toronto

Library of Congress Cataloging-in-Publication data

Walters, James.
    Aircraft accident analysis: final reports / James M. Walters, Robert L. Sumwalt III.
        p.   cm.
    Includes index.
    ISBN 0-07-135149-3
    1. Aircraft accidents—Investigation—United States.   I. Sumwalt, Walter.
    II. Title.
TL553.5.W34   2000
363.12'465'0973—dc21

                                          99-059939
                                              CIP

McGraw-Hill

A Division of The McGraw·Hill Companies

  2 3 4 5 6 7 8 9 0 DOC/DOC   0 9 8 7 6 5 4 3 2 1 0

ISBN 0-07-135149-3

The sponsoring editor for this book was Shelley Ingram Carr, the editing supervisor was Sally Glover, and the production supervisor was Pamela Pelton. It was set in Garamond per the Gen1AV1 design by Joanne Morbit of McGraw-Hill's desktop publishing department, Hightstown, N.J.

Printed and bound by R. R. Donnelley & Sons Company.

McGraw-Hill books are available at special quantity discounts to use as premiums and sales promotions, or for use in corporate training programs. For more information, please write to the Director of Special Sales, Professional Publishing, McGraw-Hill, Two Penn Plaza, New York, NY 10121-2298. Or contact your local bookstore.

 This book is printed on recycled, acid-free paper containing a minimum of 50% recycled, de-inked fiber.

Dedication

This book is dedicated to all those who have lost their lives in aircraft mishaps, and to those investigators whose relentless quest for truth helps safeguard our skies.

Contents

6. A pilot's nightmare: The inflight fire of FedEx 1406   131

7. Contact approach: A close call for Delta 554   151

Part two: Regional airline accidents

12. Dangerous misconceptions: The legacy of Comair 3272   253

Part three: Military accidents

13. A missed approach: The fatal flight of Commerce Secretary Ron Brown   277

14. A lack of teamwork: HAVOC 58 impacts Sleeping Indian Mountain   299

15. Have Blue: An airshow to remember   319

Part four: General aviation accidents

16. Shattered dreams: A record-setting flight gone awry   337

Abbreviations

AAIB	Air Accidents Investigation Branch (U.K.)
AC	Advisory Circular
ACARS	Aircraft Communications Addressing and Reporting System
ACO	Aircraft Certification Office
AD	Airworthiness Directive
ADC	Air Data Computer
ADI	Attitude Direction Indicator
ADF	Automatic Direction Finder
ADM	Aeronautical Decision Making
AEW	Airborne Early Warning (U.S. Air Force)
AFB	Air Force Base
AFD	Airfield Database (U.S. Air Force)
AFI	Air Force Instruction
AFIP	Armed Forces Institute of Pathology
AFM	Aircraft Flight Manual
AFFSA	Air Force Flight Standards Agency
AGL	Above Ground Level
AHARS	Attitude/Heading Reference System
AIA	Aviation Insurance Association
AIB	Accident Investigation Board (U.S. Air Force)
AIC	Accident Investigation Commission (Peru)
AIM	Aeronautical Information Manual
AIRMET	Airman's Meteorological Information
ALPA	Air Line Pilots Association
AMC	Air Mobility Command (U.S. Air Force)
AME	Aviation Medical Examiner
AOA	Angle Of Attack
AOPA	Aircraft Owners and Pilots Association
APA	Allied Pilots Association
APU	Auxiliary Power Unit
AQP	Advanced Qualification Program

ARAC	Aviation Rulemaking Advisory Committee
ARFF	Airport Rescue and Fire Fighting
ARTCC	Air Route Traffic Control Center
AS	Airlift Squadron (U.S. Air Force)
ASR	Airport Suitability Report (U.S. Air Force)
ASAP	Airline Safety Action Program
ASRS	Aviation Safety Reporting System
ATC	Air Traffic Control
ATIS	Automatic Terminal Information Service
AW	Airlift Wing (U.S. Air Force)
BAA	Bilateral Airworthiness Agreement
BDHI	Bearing Distance Heading Indicator
CAMI	Civil Aeromedical Institute
CAS	Calibrated Airspeed
CDU	Control Display Unit
CERAP	Center and Radar Approach Control
CERAP	Center and Radar Approach Control
CFI	Certified Flight Instructor
CFIT	Controlled Flight Into Terrain
CFR	Code of Federal Regulations
CFR	Crash-Fire Rescue
COMAT	Company-Owned Material
CPI	Crash Position Indicator (U.S. Air Force)
CRM	Crew Resource Management
CTA	Centro Tecnico Aerospacial (Brazil)
CTAF	Common Traffic Advisory Frequency
CURV	Cable controlled Underwater Recovery Vehicle
CVR	Cockpit Voice Recorder
DA	Decision Altitude
DER	Designated Engineering Representative
DGAC	Director General of Civil Aeronautics (Dominican Republic)
DGAT	Director General Air Transport (Peru)
DME	Distance Measuring Equipment
DOT	Department of Transportation
DOD	Department of Defense
DV	Distinguished Visitor (U.S. Air Force)
EAA	Experimental Aircraft Association
EGPWS	Enhanced Ground Proximity Warning System
EICAS	Engine Indicating and Crew Alerting System
ELT	Emergency Locator Transmitter
EMI	Electromagnetic Interference

EPR	Engine Pressure Ratio
FAA	Federal Aviation Administration
FAR	Federal Aviation Regulations
FBO	Fixed-Base Operator
FDR	Flight Data Recorder
FEF	Final Evaluation Flight
FL	Flight Level
FLIP	Flight Information Publication (U.S. Air Force)
FMC	Flight Management Computer
FMS	Flight Management System
FOQA	Flight Operations Quality Assurance
FSDO	Flight Standards District Office
FSHB	Flight Standards Handbook Bulletin
FSIB	Flight Standards Information Bulletin
FSM	Flight Standards Manual
FSS	Flight Service Station
GDSS	Global Decision Support System (U.S. Air Force)
GPS	Global Positioning System
GPWS	Ground Proximity Warning System
HAT	Height Above Touchdown
HIRF	High Intensity Radiation Field
HSI	Horizontal Situation Indicator
ICAO	International Civil Aviation Organization
IFR	Instrument Flight Rules
ILS	Instrument Landing System
INS	Inertial Navigation System
IRT	Icing Research Tunnel
IRU	Inertial Reference Unit
ISA	Integrated Servo Actuator
JA/AAT	Joint Airborne/Air Transportability Training (U.S. Air Force)
JIAA	Junta Investigadora de Accidentes Aéreos (Dominican Republic)
KCAB	Korean Civil Aviation Bureau
LCO	Limit Cycle Oscillations
LLWAS	Low Level Windshear Alert System
MAP	Missed Approach Point
MDA	Minimum Descent Altitude
METRO	Meteorological Service (U.S. Air Force)
MMEL	Master Minimum Equipment List
MOU	Memorandum Of Understanding
MSAW	Minimum Safe Altitude Warning system
MSL	Mean Sea Level

MV	Monovision (contact lens)
NACA	National Advisory Council on Aeronautics
NASA	National Aeronautics and Space Administration
NASIP	National Aviation Safety Inspection Program
NATCA	National Air Traffic Controllers Association
NCAR	National Center for Atmospheric Research
NDB	Nondirectional Radio Beacon
NFPA	National Fire Protection Association
NOAA	National Oceanic and Atmospheric Administration
NPRM	Notice of Proposed Rulemaking
NRS	National Resource Specialist
NTSB	National Transportation Safety Board
NWS	National Weather Service
OG	Operational Group (U.S. Air Force)
PA	Public Address system
PCU	Power Control Unit
PF	Pilot Flying
PGI	Principal international Geographic Inspector
PIC	Pilot In Command
PIREP	Pilot Report
PMI	Principal Maintenance Inspector
PNF	Pilot Not Flying
POI	Principal Operations Inspector
PSI	Pounds per Square Inch
QFE	Altimeter setting to read height above field elevation
QNH	Altimeter setting to read mean sea level
REIL	Runway End Identification Lights
RMI	Radio Magnetic Indicator
RPM	Revolutions Per Minute
RSPA	Research and Special Programs Administration
RVR	Runway Visual Range
SAE	Society of Automotive Engineers
SAAM	Special Assignment Airlift Mission (U.S. Air Force)
SIGMET	Significant Meteorological Information
SLD	Supercooled Large Droplet
SPAAT	Skin Penetrator Agent Application Tool
SUPSALV	Supervisor of Salvage (U.S. Navy)
TAWS	Terrain Awareness and Warning Systems
TBO	Time Between Overhaul
TC	Type Certificate
TCAS	Traffic Alerting and Collision Avoidance System

TERPS	Terminal Instrument Approach Procedures
THF	Tetrahydrofuran
UIUC	University of Illinois at Urbana/Champaign
USAFE	U.S. Air Force Europe
VASI	Visual Approach Slope Indicator
VDP	Visual Descent Point
VFR	Visual Flight Rules
VMS	Vertical Motion Simulator
VNAV	Vertical Navigation
VOR	Very high frequency Omnidirectional radio Range
VSI	Vertical Speed Indicator

Preface

What this book is about

In explaining what this book is about, perhaps it is best to first say what the book is not about. This is not just another treatise that bemoans that aviation is unsafe, or that this country's government oversight of air carriers is at dangerously low levels. Books of that sort seldom do more than stir up false public perceptions and create unnecessary fear of flying amongst thousands of weary travelers.

As strange as it may seem for a book dealing with aircraft accidents, the authors are firmly convinced that flying is safe. We recognize that literally millions of flights are successfully completed each year in U.S. commercial, general, and military aviation. There are many checks and balances, some operational, others regulatory, to ensure that the system remains safe. Unfortunately, there are also those rare occasions when hazardous conditions and human errors blend into a lethal mix. When that happens, accident investigators across the country are quickly launched to participate in the investigation, which is not only a fascinating series of events aimed at solving a complex puzzle, but is often an interesting story within itself.

What you will read about in this book are selected accidents—mishaps where the system failed and a catastrophic loss of life or injury occurred. There is an old saying about a person's opinion—everyone has one. The authors are no exception. But we have made a conscious effort to spare the reader our personal opinions, instead basing the information contained throughout this book strictly on the facts. The analysis and conclusions described in each chapter were decided upon not by us, but by the relevant accident investigation authorities, and only then after thousands of hours of difficult deliberation. We are simply the conduit that allows a flow of sometimes hard-to-find information, hopefully providing it to the reader in an interesting, factual, and accurate way.

The authors are well versed in aircraft accident investigation and have personally been involved in many high-profile investigations of the past two decades. We have completed courses at the National Transportation Safety Board, Transportation Safety Board of Canada, Embry-Riddle Aeronautical University, University of Southern California, the U.S. Department of Transportation's Transportation Safety Institute, and the Air Line Pilots Association. We have worked within the system, "kicking the tin," participating in Safety Board hearings, and even providing congressional testimony when necessary. We realize that as tragic as past accidents are, the lessons learned from those disasters must not be forgotten. In order to avert future accidents, all causes of a mishap must be vigorously pursued so that preventive measures can be found and implemented.

A testimony of our commitment to aviation safety is perhaps best demonstrated by what we do for our "day jobs." Both authors are captains for two of the largest airlines in the world. Each day, we entrust ourselves and our precious passengers to operate in a transportation system that we know to be very, very safe.

The irony is that some of the most effective safety enhancements in place today are the result of lessons learned from previous accidents. Philosopher George Santayana once stated, "Those who cannot remember the past are condemned to repeat it." If you are a pilot, then perhaps the accidents that are reviewed in this book will someday help you to avoid repeating the tragic misfortunes of someone else's past. The same applies if you design airplanes, maintain them on the hangar floor, work as an air traffic controller, or have some other involvement in the aviation industry. Perhaps something in this book will help you to avoid duplicating some aspect of a previous accident. When it comes to aircraft accidents, clearly it must be our goal, as Mr. Santayana aptly put it, to avoid repeating the past.

So that is what this book is about. We hope to cast an informative and factual light on aircraft accidents so that the aviation community can better understand what happened, why, and what has been done about it. It is only through this understanding that we can hope to avoid repeating the same mistakes.

We wish you interesting and informative reading, and safe flight operations!

Jim Walters
Robert Sumwalt
January, 2000

Introduction

The investigative process

It's been said that everyone loves a mystery. And there is perhaps no greater puzzle than that faced by today's aviation accident investigator. The various methodologies used to unravel the complex events surrounding the mishap can be fascinating. Between the covers of this book, the reader will get a feel for many of these investigative techniques, and will realize why accident investigation has been called an art as much as it is a science. Before launching into your own "reading investigation" of each chapter, the authors have decided to divulge a few tricks of the trade.

In the United States, the National Transportation Safety Board (NTSB) has the ultimate jurisdiction over any aircraft accident investigation. In some cases, typically general aviation accidents involving smaller airplanes, the NTSB often delegates the field investigation to the Federal Aviation Administration (FAA). However, in these cases the NTSB still retains final responsibility for determining the accident's cause and issuing appropriate safety recommendations.

The NTSB has limited staff and funding and operates according to the "party" system. Party status is generally extended only to those organizations that are familiar with the process and bring needed technical expertise and value to the investigation. For example, the FAA is always granted party status because the air traffic control system, certification of aircraft and aviators, and enforcement of the federal aviation regulations fall under the FAA's purview, and to some extent, these issues are always considered during the course of an investigation. The aircraft and engine manufacturers are typically parties to the investigation because they have the most knowledge about and experience with their products. If the accident involves an airline, you can expect the airline will be a party, along with the pilots' and flight attendants' unions. Why is this? Because the NTSB has learned over the years that these different organizations each bring unique knowledge and perspective to the investigative process. Who better knows that particular airline's procedures, policies, and accepted practices than the pilots who fly those airplanes and the company for whom they are employed?

As Jim Hall, Chairman of the NTSB, stated recently to a gathering of aviation attorneys, "The underlying premise of the party system is a strong one—that everyone has an overriding interest in safety and that everyone wants to find out what happened so that steps can be taken to ensure such an accident is not repeated." The system works. As Chairman Hall explained, "The party participants—be they air carriers, manufacturers, pilot organizations, emergency response providers, suppliers or maintenance providers—all have at one time or another provided the NTSB with the technical depth of knowledge we have needed to determine the probable cause of a transportation disaster..."

On the site

What does an investigator do upon reaching an accident site? "When you first go to the scene of an aircraft accident, the wreckage tends to draw you like a magnet right to it," explains a NTSB instructor. "That's one of the things that you shouldn't allow to happen. Notice the trees, the buildings, the weather, all of the things that encompass the environment surrounding the accident site. That way, when you finally do look at the wreckage, you'll get more meaningful information."

Upon first arriving on scene, one of the first things that is recommended is for the investigator to perform a "walk through" of the wreckage site, to get a general feel for the situation. Documenting the wreckage should be the next order of priority, by photography and precise diagramming. General layout of the destroyed aircraft provides important clues as to how and why the pieces ended up where they did. To the experienced investigator, ground scars can tell a fascinating story of the final flight path and "kinematics," or crash dynamics, of how the aircraft initially struck terrain. The distance between propeller slash marks can be measured, and if the prop's RPM is known, the aircraft's ground speed at impact can be determined. Conversely, if ground speed is known, the prop RPM at impact can be calculated (see Chapter 14).

Trees can provide additional clues. In a heavily wooded area, their damage patterns can indicate the flight path angle of descent and aircraft bank angle. This information coupled with the general wreckage distribution pattern begins to tell the story of the last moments of flight. Generally, wreckage that resulted from a steep angle of impact remains relatively close together (see Chapter 1 and Chapter 5). If pieces of wing or tail sections are found far from the main wreckage, investigators will consider the possibility of an inflight breakup (see Chapter 15).

A day or two after the accident, dead leaves on an otherwise green tree may be an indication of fuel spilling onto the leaves during impact. If there is a question of whether or not fuel was onboard at the time of impact, this may be a good clue. Questions about fuel type can be answered by having the leaves tested in a lab to check their flash point and vapor pressure. Each fuel type has its unique signature for these items.

All of these are only general rules of thumb, however, and to ensure validity they must be carefully weighed with other evidence. "Until you get the big picture, a lot of things can be masked. Get the complete picture and don't jump to conclusions," warns the NTSB instructor.

In the case of a suspected inflight breakup, the investigator would expect to find evidence of wing or tail overstressing. With positive G overstressing, the wings are forced upward relative to the fuselage. The upper spar cap may show signs of "compression stress" due to this leveraged upward bending. Likewise, the skin of the wing's upper surface may be wrinkled. Conversely, the lower spar cap may show signs of stretching, or "tensile stress," as it is called in the accident investigator's vernacular.

Ice that accumulates on the wing and tail leading edges during flight can have a devastating effect on the performance of the aircraft (see Chapter 12). It is usually difficult to find evidence of inflight icing in wreckage because it will probably have melted by the time the investigator arrives on scene. But if these surfaces are not damaged by fire, careful inspection can still provide clues. If the leading edge of an airfoil is clean of dirt, grime, and debris, and the rest of the wing has telltale signs of impact dirt and soot, this could indicate that ice was covering the wing's leading edge at impact, then melted off.

Propeller damage may support evidence of an engine developing power (see Chapter 17). Significant degrees of propeller twisting, prop leading edge damage, and scratch marks across the propeller's surface can be indicators of engine power. This is one of the more difficult areas to analyze, however, and is usually left to the propeller experts. Again, there are few hard-and-fast rules, so the investigator must weigh all of the evidence together.

Aircraft collisions, whether in midair or on the ground, will leave telltale paint transfer marks and scratch marks on one or both aircraft. By measuring the angles of these marks relative to the aircraft's longitudinal axis, the investigator uses basic trigonometric functions to determine collision angle. This information can be critical in determining whether the pilots could have seen each other, or whether obstructions inside or outside the aircraft limited their visibility (see Chapter 11).

When an investigator finds wreckage that is burned, there's always the issue of whether the fire was the cause or the result of the accident. With an inflight fire the aircraft may have dropped debris and hardware in the last moments of flight. Therefore, when inflight fire is suspected, the investigator will search the areas back along the final flight path for supporting evidence. The investigator should piece aircraft parts back together to compare burn marks. Finding one part scorched but the adjoining part untouched by fire suggests the obvious: the parts had separated before fire eruption. The source of an inflight fire can be difficult to ascertain, even when there seems to be ample evidence (see Chapters 5 and 6). But painstaking reconstructive efforts usually solve the puzzle.

Even with charred or fragmented cockpit instruments, it is sometimes possible to determine the indicator's position at impact. The needle (if an analog, or "round dial" cockpit) of an instrument may momentarily strike the instrument face at impact, then spring back rapidly and strike the back side of the glass cover. This spring action of the needle may leave a slight impression in the instrument's face, or it may leave a subtle deposit of luminous paint on the instrument's glass. These trademarks may be detected with an ultraviolet light (see Chapter 13).

Digital aircraft utilizing "glass cockpit" technology present additional challenges, and opportunities, to the investigator. No longer can wrecked instruments be coaxed into sharing their secrets—only computer chips remain. But those circuits sometimes hold a wealth of information for the knowledgeable investigator (see Chapter 2).

By looking at a light bulb's filament it may be possible to determine if the bulb was illuminated at impact. The filament of an illuminated bulb is warm and relatively flexible. At impact this filament tends to stretch and become elongated. Conversely, an unlit bulb's filament will be relatively cold and brittle and may shatter upon impact.

Behind the scenes

When the on-site investigation is completed, there is still potential evidence to be gathered. ATC radio transmissions and radar presentations are recorded, and those tapes are available to the investigation if requested. Radar information provides a "road map" of the position of the accident aircraft from takeoff to landing, and audio information can be extremely helpful in verifying and timing cockpit voice recorder (CVR) and flight data recorder (FDR) data. Weather reports need to be collected and maintenance records should be carefully reviewed.

Aircraft components may need to be sent off-site for examination. A portion of the wing spar could be sent to a metallurgy lab to confirm the suspected mode of failure. The powerplant may need tear down and inspection by the manufacturer. For aircraft so equipped, the CVR and FDR must be sent to the investigative headquarters for readout.

Witness interviews should be conducted as soon after the accident as possible. If too much time elapses, the witnesses' thoughts may be affected by news media reports and other sources. Additionally, with the passage of time the witness can be affected by "closure theory," where the witness's mind subconsciously tries to "fill in the blanks" to help reconcile the traumatic event that he or she witnessed. For instance, witnesses of inflight breakups often report hearing an explosion, followed by the wings falling off. In reality, this loud popping sound is usually caused by the sudden release of the wing's potential energy when the wing spar snaps. However, the witness's mind subconsciously links the loud popping noise and the wing separation with an inflight explosion. The person's mind may then quietly add other elements associated with explosions, such as fire and smoke. Analyzing witness statements can become one of the more difficult activities associated with the investigation—frequently there are as many versions of an observed incident as there are observers!

The human performance portion of the investigation will focus on items like interviewing the last person who saw the pilot before flight. It is recommended that the investigator start with the moment of the accident and work backwards, conducting a seventy-two-hour history of the crew to learn their activities prior to the mishap. But this historical "autopsy" should only be the very beginning, not the end of the human performance investigation. It will never be enough just to simply know that someone made a mistake—the real issues are why they did it, could it happen to someone else, and how can it be prevented in the future?

You're off on your own

We've provided you with a broad look at some of the techniques used to investigate aircraft accidents. Now, as you read each of the chapters in this book, see if you can apply some of what you have learned. Good luck with your sleuthing!

Part One

Major Air Carrier Accidents

1

The longest investigation in U.S. history

USAir flight 427

Operator: USAir

Aircraft Type: Boeing 737-300

Location: Near Aliquippa, Pennsylvania

Date: September 8, 1994

In the summer of 1992, one of this book's authors attended the National Transportation Safety Board's (NTSB) Aircraft Accident Investigation School. During the first day of class, a seasoned investigator tried to prepare his students for something that words can not adequately describe. "When you first walk onto an accident site, it appears to be a big jigsaw puzzle—a real jumbled-up mess!"

Two years later, the author—by then a trained accident investigator—would enter his first full-scale airline accident investigation. It was a big one, too. An airliner had just gone down outside of Pittsburgh. Indeed, it would be a jumbled-up mess. Indeed the investigation would be like a big jigsaw puzzle. Sometimes the pieces fit, but many times they would not. During the next four-and-a-half years, investigators would log more hours solving this mystery than any other in the NTSB's thirty-two year history. The investigation was spectacular and methodical, drawing on resources and personnel from all corners of the globe. In their probing, investigators uncovered a previously unknown flight control failure mode of the Boeing 737. This latent condition was not only responsible for the downing of USAir 427, but also would provide the key to uncloaking an enigma surrounding another B-737 accident that had occurred years earlier.

Flight history and background

After long, sultry summer days, welcome are the first cool, clear breezes that signal the imminent arrival of fall. Snapshot September 8, 1994: beautifully clear blue skies, unlimited visibility, cool temperatures, and light winds.

Just after noon that day, two USAir pilots and three flight attendants reported for duty at the Jacksonville International Airport, Florida. The crew was in a good mood; everyone seemed to get along well. The evening before as they deplaned for their thirteen-hour Jacksonville layover, the crew sang "Happy Birthday" to one of their flight attendants. Today was like a Friday for this crew, as it was scheduled to be the end of their three-day trip sequence.

Flying N513US, a B-737-300 (see Fig. 1-1) that was delivered to USAir seven years earlier, the crew flew USAir flight 1181 from Jacksonville to Charlotte, N.C., and then to Chicago O'Hare International Airport (ORD). They landed in Chicago just as the afternoon rush hour was gearing into full swing.

Almost an hour later they departed ORD, now flying as USAir 427 bound for Pittsburgh International Airport (PIT). Flying time was scheduled to be just under one hour. The forty-five year old captain was performing radio communications and other Pilot Not Flying (PNF) duties, while the thirty-eight year old first officer was performing the Pilot Flying (PF) duties. Between the two of them, there was a wealth of flying experience. Their combined flight times exceeded 21,000 hours, with over 7,500 hours in the B-737. The captain was described as being meticulous, very proficient and professional, attentive to detail, and well liked. The first officer was likewise considered an outstanding first officer who exhibited exceptional piloting skills.

As the aircraft approached its destination, Pittsburgh Approach Control directed USAir 427 to turn right to a heading of 160 degrees and advised that they would receive radar vectors for the final approach course to Runway 28R. As the air traffic controller was planning to sequence flight 427 behind Delta 1083, a Boeing 727, he further advised USAir 427 to reduce speed to 210 knots. The flight was then

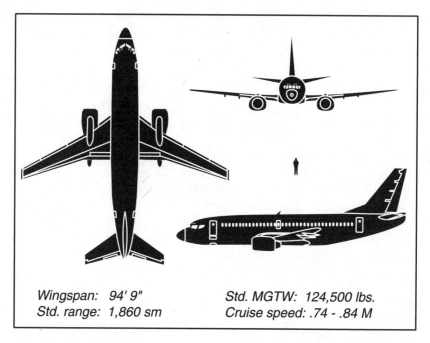

Wingspan: 94' 9"   Std. MGTW: 124,500 lbs.
Std. range: 1,860 sm   Cruise speed: .74 - .84 M

1-1 *Boeing 737-300.*

cleared to descend to 6,000 feet mean sea level (msl). The time was 6:58 P.M.

About a minute-and-a-half later, air traffic control (ATC) instructed Delta 1083 to turn left to 130 degrees and reduce speed to 190 knots. Still intending for USAir 427 to follow the approximate flightpath of Delta, the controller then assigned USAir 427 a heading of 140 degrees and an airspeed of 190 knots. The USAir crew selected flaps to the "Flaps one" setting to allow the aircraft to slow.

At 7:00:43, ATC instructed Delta 1083 to turn left to a   heading of 100 degrees. A minute-and-a-half later, Pittsburgh Approach stated, "USAir 427, turn left heading one zero zero. Traffic will be one to two o'clock, six miles, northbound, [a] Jetstream climbing out of thirty-three [hundred feet] for five thousand [feet]."

"We're looking for the traffic, turning to one zero zero, USAir four twenty seven," replied the captain. The 737 was flying at 190 knots, level at 6,000 feet with the autopilot engaged. As it approached the ATC-assigned heading of 100 degrees, the aircraft began smoothly rolling out of the left bank toward a wings-level attitude. At this time, about 7:02:53, the first officer remarked to the captain, "Oh, ya, I see zuh Jetstream." As he completed that statement, the cockpit voice recorder (CVR) recorded three rapid thumps.

"Sheeez," exclaimed the captain.

"Zuh," uttered the first officer, as if he was beginning to say something, when he was suddenly surprised.

During this time, the aircraft's indicated airspeed quickly fluctuated from 190 knots to 193 knots and then back to 191 knots. The left bank angle, which previously had been smoothly rolling out on the assigned heading, increased from about seven degrees to more than twenty degrees in the next two seconds. The CVR recorded an additional thump, two "clickety click" sounds, the sound of increasing engine noise, and the sound of the captain inhaling and exhaling quickly one time.

About 7:02:59, the rapid roll to the left was arrested, and the airplane began to briefly roll right towards a wings-level attitude. But also about this time, the airplane's heading, which had been moving left steadily towards the ATC-assigned heading of 100 degrees, began to move left at a more rapid rate, passing through the 100-degree heading.

"Whoa," stated the captain. The CVR, with enhanced audio recording capability due to the crew's wearing of boom microphones, detected the first officer grunting softly.

The arresting of the left rolling movement was short-lived. By just after 7:03:00, the airplane had begun to roll rapidly back to the left again; the airplane's heading had moved left through about 089 degrees and was continuing to move left at a rate of at least five degrees per second. Between about 7:03:01 and about 7:03:04, the CVR recorded the first officer again grunting, this time more loudly. By now the airplane's left bank angle had increased to about forty-three degrees, and the airplane had begun to descend from its assigned altitude of 6,000 feet msl. There were several rapid fluctuations in the rolling movement. The first officer disconnected the autopilot. The left roll continued. By 7:03:06, the left bank angle exceeded seventy degrees.

At 7:03:07.5, the aircraft began buffeting as the airflow over the wings was disrupted due to excessive aerodynamic angle-of-attack. In short, the wing was stalling. "What the hell is this?" demanded the captain.

The stickshaker, the 737's stall warning system, began rapidly vibrating the control wheel. Within half a second, the altitude alerter sounded to alert the crew that the aircraft had deviated more than 300 feet from its assigned altitude. Within the next second, the onboard Traffic Alerting and Collision Avoidance System (TCAS) detected a possible traffic conflict with the Jetstream that was climbing to 5,000 feet, and announced an advisory message, "Traffic, traffic." About this time, the approach controller noticed that USAir 427's altitude readout on his ATC radar screen indicated that the 737 was descending through 5,300 feet.

"USAir 427 maintain 6,000, over," stated the controller.

"Four twenty seven emergency!" responded the captain.

Within seconds, the aircraft would be pointed nearly vertically towards the ground. Between 7:03:18.1 and 7:03:19.7, the CVR recorded the captain commanding "pull...pull...pull," perhaps in a desperate attempt for the first officer to pull the nose up. By that point, however, there was nothing that could be done.

The CVR stopped recording at 7:03:22.8.

The airplane impacted hilly, wooded terrain near Aliquippa, Pennsylvania, approximately six miles northwest of PIT, at an elevation of about 930 feet msl. The ensuing post-impact fire burned for approximately five hours, but the wreckage continued to smolder for days. There were 132 fatalities.

The investigation and findings

A full NTSB go-team arrived in Pittsburgh early the next morning. Although the weather the previous day had been perfectly clear, by the time investigators began their initial probe a massive downpour had thoroughly drenched the crash scene. Mud, debris, and carnage were everywhere.

The aircraft's primary impact point was in a densely wooded area on an up-sloping hillside alongside a dirt road cut through the trees. The wreckage was severely fragmented, crushed and burned, and most was located within a 350-foot radius of the main impact crater. It was computed that the aircraft impacted terrain at about an eighty-degree nose-down attitude, in a slight roll to the left.

Examination of the spoiler control surfaces and actuators revealed that all wing spoilers were in the retracted position at impact, with no evidence of preimpact failure. Inspection of the leading-edge flaps and slats revealed they were extended symmetrically at impact, with no evidence of any precrash malfunctions. Likewise, there was no evidence of structural fatigue or preimpact fire on the trailing edge flaps or flap tracks.

The Flight Data Recorder (FDR) indicated that the engines were operating normally and symmetrically until ground impact, and physical inspection of what remained of the engines supported that finding. Examination of the engine thrust reversers indicated that the left engine reverser locking actuators were in the stowed position at impact. However, the actuators for the right engine were discovered in the extended position. Two days into the investigation, perhaps this was the break that investigators needed. It wouldn't be that easy, though. Subsequent X-ray inspection and disassembly of the four reverser locking actuators revealed that all four actuator pistons were in the stowed position with locking keys engaged at impact. The anomaly noted on the right engine was determined to be the result of impact forces.

The airplane's tail section was located in an inverted position. The horizontal stabilizers and elevators remained attached to the tail (see Fig. 1-2). Flight control continuity was established within the tail section, with an elevator position of about fourteen degrees nose-up. The vertical stabilizer and rudder were located next to the tail section, and the aft rudder control quadrant was found attached to its mounting brackets. The main rudder power control unit (PCU) displayed a bend in the actuator rod that was consistent with a rudder position of about two degrees to the right (airplane nose right). If this was true, then why was the aircraft turning to the left, investigators wondered.

Only a few fragments of the cockpit were recovered. A notable find was the airspeed indicator, frozen at 264 knots or 303 miles per hour (mph).

A ground and helicopter search was conducted of the aircraft's final flightpath to look for additional components, but none were found. Because the airplane impacted at such a high speed, the NTSB was concerned that important aircraft pieces may have penetrated the topsoil into harder clay and might not be easily located and recovered.

1-2 *Tail wreckage of USAir 427. . .* Courtesy Capt. John Cox, ALPA

Using a ground-penetrating radar borrowed from the U.S. Bureau of Mines, additional pieces of wreckage were discovered six feet under ground.

Before removal from the accident site, the wreckage was thoroughly examined, components were identified and photographed, and critical measurements were recorded. After the wreckage was documented and decontaminated, it was taken to a hangar at PIT for further examination and a two-dimensional reconstruction.

The purpose of the reconstruction was to determine whether a control cable failure, bird (or other airborne object) strike , floor beam failure, or in-flight explosion was involved in the accident. The process involved using Boeing drawings to identify aircraft pieces; once identified, workers would then lay them out somewhat to scale in their relative positions on the hangar floor (see Fig. 5-3, a photograph of a similar reconstructive effort involving Valu-Jet flight 592). The NTSB enlisted the aid of the British Air Accidents Investigation Branch (AAIB), because of that agency's experience with the reconstruction of the in-flight explosion of Pan Am 103 near Lockerbie, Scotland, some six years earlier. Although reconstruction efforts were severely hampered because much of the airplane was destroyed or heavily damaged, the AAIB could find no evidence of any preimpact explosion. Additionally, explosive experts from the FAA (Federal Aviation Administration) and the FBI (Federal Bureau of Investigation) examined the wreckage and formed similar opinions.

The Safety Board also examined the CVR and FDR information from the accident airplane and compared it with FDR and CVR information obtained from Pan Am 103 and other known in-flight fire, bomb, and explosion events. No similar signatures were found. All aircraft doors and hatches were accounted for, and their respective locking mechanisms provided witness marks consistent with them being closed and locked at impact. About forty percent of the rudder control cables were recovered and those remains showed no evidence of preimpact failures.

An ultraviolet light, commonly used to detect blood, was used to look for evidence of a bird strike. Examined were portions of the radome, the forward pressure bulkhead, left wing leading edge slats, cockpit flight control components, and the leading edges of the vertical and horizontal stabilizers. No evidence of a bird strike was found.

Although the CVR unit showed evidence of external and internal structural damage, the recording tape was in good condition. The quality of the recording was classified as "excellent," which by NTSB standards means that virtually all of the crew conversations could be accurately and easily understood.

The crash-protected FDR memory module unit and recording medium were intact and provided good data. Although this FDR actually recorded thirteen parameters, it was often referred to as an "eleven parameter FDR," because the Federal Aviation Regulations (FARs) that required its installation did not specifically require recording two of those thirteen parameters. Recorded parameters that were sampled at once-per-second intervals were altitude, indicated airspeed, heading, and microphone keying. Also recorded once-per-second for each engine were exhaust gas temperature (EGT), fuel flow, fan speed (N1) and compressor speed (N2). Recorded parameters that were sampled at more frequent rates were roll attitude and control column position (two times per second), pitch attitude and longitudinal acceleration (four times per second), and vertical acceleration (eight times per second).

The FDR did not record actual flight control surface positions, nor was it required to by the FARs. Without that information, however, investigators were forced to spend the next several years conducting complex computer simulations and a mathematical study known as a "kinematic analysis" to derive their best estimates of where and when the control surfaces moved. And after all of that, the real question remained: how did the controls get to those positions?

Years later, as the investigation was concluding, NTSB Chairman Jim Hall confessed, "I wish it had not taken us four-and-a half years to get to this point, but the complexity of this investigation, coupled with the appalling lack of flight data recorder information, necessitated a long, comprehensive investigation."

The roll event

It was clear from witness statements and FDR information that the aircraft rolled to the left during its fatal plunge to earth. Investigators considered various scenarios that could have produced the rapid leftward rolling moment. Slowly, one by one, they conclusively ruled out the possibility of asymmetrical engine thrust reverser deployment, asymmetrical spoiler/aileron activation, transient electronic signals causing uncommanded flight control movements, yaw damper malfunctions, and a rudder cable pull or break. While is it true that each of these events could have caused a rapid rolling moment, investigators determined that none of them could have matched the heading change and acceleration curves that were documented on the FDR.

Wake turbulence

The NTSB obtained radar data from Pittsburgh Approach Control and plotted the positions of USAir 427 and Delta 1083 to see if wake turbulence from the Delta 727 could have somehow triggered the upset. The radar data showed that Delta 1083 was descending through 6,300 feet msl on a 100-degree heading when it passed the approximate location where the initial upset of USAir 427 subsequently occurred. USAir reached that location about sixty-nine seconds after Delta 1083, and the two airplanes were about four nautical miles apart. NTSB and NASA aerodynamics experts created a wake turbulence model and concluded that under the atmospheric conditions that evening, the wake vortices would have descended approximately 300 to 500 feet per minute. This placed them at the same point in space that USAir 427 subsequently flew through, and at the point where the upset initially began. From this, investigators drew the conclusion that USAir 427 did encounter wake turbulence from Delta 1083.

The NTSB was aware that wake turbulence had led to three air carrier accidents between 1964 and 1972. However, each of these aircraft were operating at low altitudes during takeoff and landings, unlike USAir 427, which was flying relatively fast and at higher altitude. After the 1972 wake turbulence accident, ATC separation standards were increased and since then, there have been no fatal wake turbulence-related air carrier accidents. So, investigators were baffled at how wake turbulence could have played a role in this accident.

In an effort to learn more about wake turbulence, the NTSB conducted a series of wake turbulence tests near Atlantic City, New Jersey, in the autumn of 1995. These tests used a specially instrumented Boeing 737 to fly through the wake of a 727, whose wake was marked by smoke generators mounted on each wingtip. Care was taken to load each aircraft to the approximate weights of Delta 1083 and USAir 427, as well as to conduct the tests during atmospheric conditions similar to those that existed when the accident occurred.

During the tests, the 737 penetrated the 727's wake vortex cores more than 150 times, from different intercept angles, flight attitudes, and distances ranging from 2 to 4.2 miles. Each of these penetrations was recorded with a video camera mounted in the 737 cockpit and a camera with a wide-angle lens mounted on the tip of the vertical stabilizer. Also used was a special FDR that sampled specific parameters twenty times per second.

Various pieces of audio recording gear were placed throughout the aircraft to see if any of the mysterious thumps and other puzzling sounds on 427's CVR could be replicated. The NTSB then performed a sound spectrum analysis of several curious sounds. As it turned out, the three thumps on the CVR had the same audio characteristics as those heard during the wake turbulence tests, as the main aircraft fuselage passed through the wake's core.

The test pilot participating in these tests remarked that the wake turbulence had varying effects on the 737 flight handling characteristics. He also stated that the effects usually lasted only a few seconds and did not result in a loss of control or require extreme or aggressive flight control inputs to counteract. Review of the test data did not reveal any instances in which the wake vortex encounter produced a heading change resembling that recorded on flight 427's FDR. It was further determined that in most of the wake tests, once the aircraft entered the vortex, the natural tendency was for the aircraft to be ejected from the wake's effect. Based on all of these data, the NTSB determined that flight 427 would not have remained in the vortex long enough to have produced the heading change and bank angles that occurred after 7:03:00 P.M. Investigators believed that there had to be more to it than just wake turbulence.

Left rudder deflection

While several factors were being ruled out as a possible cause of the 737's roll, one theme continued to ring constant. The available FDR data, although limited, matched computer simulations that indicated the airplane had likely experienced a sustained full left deflection of the rudder. The focus of the investigation then narrowed to determine what events could have resulted in that rudder movement. Investigators were convinced that it had to be one of two things: a mechanical rudder system anomaly or flightcrew action.

Boeing 737 rudder system

During normal in-flight operation a pilot turns the aircraft using the control wheel, which banks (rolls) the aircraft by deflecting ailerons on the back of each wing. The 737 also uses wing-mounted spoilers to augment roll control. As the ailerons (and spoilers) deflect and the aircraft begins to turn, it will also tend to yaw (turn) opposite the commanded direction, due to increased drag on the wing that is being lowered and increased lift on the wing that is being raised. This yawing tendency is appropriately known as "adverse yaw." The rudder is used to counteract the adverse yaw, and is controlled by the pilot through floor-mounted rudder pedals. As a pilot rolls into a turn by turning the control wheel, he/she should also apply an appropriate amount of rudder to keep the aircraft from yawing too much into or out of the turn (see Fig. 1-3).

The rudder is also normally used by pilots during crosswind takeoffs and landings to keep the aircraft tracking straight down the runway centerline. It is also used during abnormal situations to counter asymmetric turning tendencies such with an engine failure or asymmetric flap condition.

In the 737, each pilot has a pair of rudder pedals. These pedals are connected by cable to the tail section of the aircraft, where they are joined to the main rudder Power Control Unit (PCU) and the standby rudder PCU, located in the aft portion of the vertical stabilizer. The

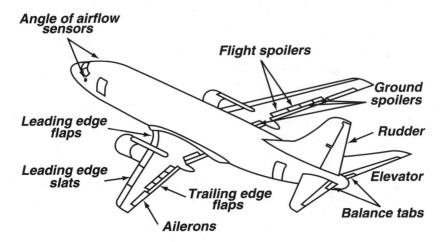

Angle of airflow sensors

Flight spoilers

Ground spoilers

Leading edge flaps

Rudder

Leading edge slats

Elevator

Trailing edge flaps

Ailerons

Balance tabs

1-3 *Boeing 737 flight control surfaces.* . . Source: NTSB

PCUs, through various mechanical and electronic inputs, convert pilot commands (mechanical inputs through the rudder pedals and cables) into hydraulic outputs that operate the rudder. The 737-300's rudder is a single panel, and during normal operation it is actuated by a single hydraulic PCU (the main rudder PCU).

Unlike most light airplanes, in many large transport-category aircraft such as the 737, it is not possible to move the rudder without hydraulic pressure. In the event that both of the 737's primary hydraulic systems are lost ("System A" and "System B"), the "standby" hydraulic system is available to power the rudder. The standby hydraulic system is totally independent of Systems A and B, and has its own standby rudder PCU.

The NTSB conducted a review of large transport-category aircraft, including Boeing, McDonnell Douglas, Airbus, and Lockheed models, and found that the 737 is the only twin wing-mounted engine, large transport-category airplane designed with a single rudder panel and a single rudder actuator. All other large transport-category airplanes with twin wing-mounted engines were designed with a split rudder panel, multiple hydraulic actuators, or a mechanical/manual/trim tab rudder actuation system. The NTSB observed that because the 737 engines are wing-mounted, its rudder system has to be sufficiently powerful to effectively counter the significant asymmetric yaw effect of a loss of one engine.

When properly rigged and installed, the 737-300 main rudder PCU can command a maximum deflection of twenty-six degrees left or right from its neutral position, under no aerodynamic loads. As the aircraft goes faster through the air, air pressure from the slipsteam will limit the amount of available rudder deflection. For illustration, if a rudder is able to move twenty-six degrees left or right with no aerodynamic loading (such as on the ground under static conditions), that same rudder might only deflect to twenty-two degrees when the aircraft is flying at 300 knots. The maximum amount that the rudder can deflect under given flight conditions/configuration is known as the rudder's "blowdown limit."

Main rudder PCU servo valve

The major "working part" of the PCU is the servo valve, located inside the main PCU housing. This valve was designed by Boeing and manufactured to Boeing specifications by Parker Hannifin Corporation. The purpose of the servo value is to direct hydraulic pressure to the appropriate "ports," or small openings in the assembly, to move the rudder in the proper direction. This can be commanded either by the pilot (through the rudder pedals), the yaw damper (which can move the 737-300 rudder up to three degrees either side of neutral without pilot input to maintain aircraft stability while in turbulence), or from rudder trim.

The servo valve is called a "dual concentric tandem valve," meaning that there are really two valves, or slides, in one housing. The primary slide is about the size of a pencil, and has small groves and holes machined into it to direct hydraulic fluid to the rudder actuators. When the PCU receives a command to move the rudder, the primary slide moves in the appropriate direction to send hydraulic pressure that either extends or retracts the rudder actuator piston, which is directly linked to the actual rudder panel. When the rudder actuator piston is moved in the extend direction, it moves the rudder left; when it moves in the retract direction, it moves the rudder to the right.

Concentrically surrounding the primary slide is the secondary slide, and it is concentrically surrounded by the servo valve housing. Like the primary slide, the secondary slide can also move back and forth to port hydraulic pressure to move the rudder. Normally, when a command is received to move the rudder, the primary slide displaces first. The secondary slide moves only when movement of the primary slide is not sufficient to move the rudder at the commanded rate. The two slides are designed to provide approximately equal hydraulic fluid flow. The primary slide alone can provide a rudder rate of about thirty-three degrees per second, and the primary and secondary slides together can provide a rudder rate of about sixty-six degrees per second under zero aerodynamic conditions (see Fig. 1-4).

The total distance that these slides actually move is very limited. From neutral to their full extreme positions is only about 0.045 inch for the primary and secondary slides, for a combined distance of 0.090 inch. This is about the thickness of a dime. Both slides are designed so that they can move about 0.018 inch axially beyond their normal operating range, and this is known as "overtravel" capability.

These slides were designed to fit together with very tight clearances so the servo valve could be manufactured without "O-rings" to seal the slides. There is about 0.00015 inch (less than the thickness of a human hair) clearance between the outer diameter surfaces of the primary slide and the inside diameter surfaces of the secondary slide, as well as the outside diameter of the secondary slide and the inside diameter surfaces of the servo valve housing (see Fig. 1-5).

NTSB testing of the PCU and servo valve

The PCU recovered from the wreckage of USAir 427 was subjected to many tests. The yaw damper system was examined to see if a failure mode could have allowed it to exceed its three-degree authority for moving the rudder. Actual hydraulic fluid trapped in lines and actuators from the accident aircraft was examined to see if it contained contamination that could have resulted in a rudder abnormality. In

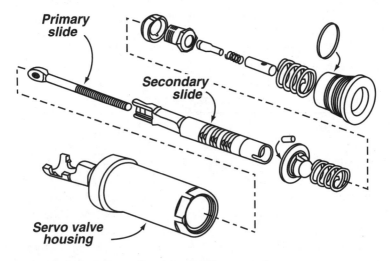

1-4 *Exploded view of rudder PCU servo valve. . . .* Soruce: NTSB

1-5 *Photograph of PCU servo valve with case cutaway to show internal components.* . . . Courtsey Jan Steenblik, <u>Air Line Pilot</u> magazine

another test, small chips of rubber, wire and hardened-steel were placed into a test servo valve to see if a jam could be introduced. As designed, most chips were sheared off and the PCU continued to operate, however, one of the larger steel chips did lodge in the servo valve. Upon disassembly, investigators found physical marks on the servo valve where the chip had lodged. No such marks were found on the servo valve from flight 427. Testing was conducted to evaluate the effects of air in the 737 hydraulic system and to check for the possibility of "silting," a process where extremely small particles build up in the servo valve and compromise the already tight clearances designed into that system. Each of these tests, while helpful for ruling things out, provided investigators with little in the way of answering why flight 427 crashed.

Independent technical advisory panel

Perhaps investigators were feeling frustrated that wherever they turned, testing provided no real clues. In January 1996, the Safety Board took the unprecedented step of forming an independent advisory panel of six government and industry experts. This group's mission was to review the work accomplished in the investigation to ensure that all issues had been fully addressed, and to propose any additional efforts that they believed were needed to ensure a complete and thorough investigation.

During their first meeting, one of the panel members stated that he had worked with a military fighter project that had used a control system PCU similar in design to the 737 main rudder PCU. He stated that an accident occurred in a very early test flight that was attributed to a jammed PCU. The investigation of that mishap revealed that the unit jammed when a sudden full-rate input caused hot hydraulic fluid to enter the cold PCU. The inner parts of the PCU thermally expanded into the cold PCU body, resulting in a jammed condition. Armed with this input, the USAir 427 crash sleuths developed a thermal shock test plan.

Thermal testing of the PCU

Boeing engineers confirmed that failure of one of the engine-driven hydraulic pumps could result in the overheating of hydraulic fluid. In August 1996, twelve thermal tests were conducted; four were on a new-production PCU and eight on the USAir 427 PCU. In October 1996, nineteen more tests were done; eight were on the new-production PCU and eleven on the PCU from flight 427.

One of the various tests was an "extreme temperature differential test." This test involved precooling the PCU to about minus 40 degrees Fahrenheit and then injecting hydraulic fluid that had been heated to about 170 degrees Fahrenheit.

The results of the August 1996 testing were surprising. The new-production PCU responded normally, but the 427 PCU showed anomalous behavior. In one of these tests, the 427 PCU stuck in the full left rudder position for about five seconds. In another, it stuck for about one second. In other tests, it exhibited slower than normal movement for the left rudder command.

The October 1996 tests yielded similar results. According to the NTSB, "Further examination of the data indicated that the servo valve secondary slide momentarily jammed to the servo valve housing and that the subsequent overtravel of the primary slide resulted in an increase in system return flow that could cause a rudder actuator reversal (travel in the direction opposite to that commanded.)"

After the thermal testing was completed, the USAir 427 PCU was disassembled and examined at Parker Hannifin. The primary slide, secondary slide, and the interior of the servo valve housing showed no evidence of damage or physical marks from jamming or binding that occurred during the thermal testing. Also noteworthy was that before disassembly, the PCU passed the functional acceptance test used by Parker Hannifin to validate PCU performance. This failure mode was elusive; although the PCU had been observed to jam in several tests, it still passed the acceptance tests and provided no witness marks.

Rudder reversals during secondary slide jams

Following the Safety Board's October 1996 thermal tests, Boeing conducted independent tests using a new-production PCU that was modified to simulate a jam of the secondary slide. These tests revealed that when the secondary slide was jammed to the servo valve housing at certain positions, the primary slide could travel beyond its intended stop (overtravel) because of bending or twisting on the PCU's internal input linkages. This deflection allowed the primary slide to move to a position where the PCU commanded the rudder in the opposite direction of the intended command (see Fig. 1-6). Eureka, we have a rudder reversal!

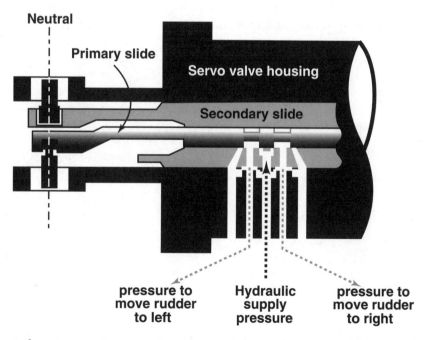

1-6 *Cutaway diagram of rudder PCU servo valve. . .* Source: NTSB

Other significant roll/yaw events

During the investigation, the NTSB gathered a listing of several 737 rudder-related or roll-related events. Investigators learned of two events where pilots who were incapacitated by seizures inadvertently pushed on a rudder pedal, causing an unexpected roll. But in many more events, there was a mechanical explanation. For example, in July 1974, a flightcrew reported that the rudder "moved full right" upon touchdown. The investigation revealed that the primary and secondary slides of the main rudder PCU had jammed together due to a foreign object that became lodged in the servo valve. In another case, in January 1993, a crew reported "binding" of the rudder pedals during their pretakeoff flight control check. The main rudder PCU was immediately shipped to Parker Hannifin for analysis. During testing, the unit exhibited rudder reversals, meaning that when a right rudder movement was commanded, the rudder would actually move left. While many of these events were insightful, there were two events in particular that the NTSB considered key to the 427 investigation.

United Air Lines flight 585

In March of 1991, NTSB investigators found themselves wrestling with a troubling Boeing 737 accident. United 585, a Boeing 737-200, crashed under mysterious circumstances while on approach to Colorado Springs, Colorado. The Board began to realize that there were striking similarities between that accident and the USAir 427 accident. The United aircraft encountered turbulence, not from wake vortices as with USAir 427, but rather from strong and gusty atmospheric winds. As the aircraft was about 1,000 feet above ground, it suddenly yawed and rolled rapidly to the right, followed by a rapid pitch down and crash. United 585 only had a five-parameter FDR, however, which severely hampered the original investigation.

When concluding 585's investigation in December 1992, the NTSB, for only the third time in their history, announced that they would be unable to positively determine a probable cause. They listed two possible causes: a malfunction with the aircraft's lateral or directional control system, or an encounter with an unusually severe atmospheric disturbance, such as a mountain rotor.

The aircraft's maintenance history revealed that in the week prior to the accident there were two rudder-related pilot "write-ups." One stated that "on departure got an abnormal input to rudder that went away. Pulled yaw damper circuit breaker." The yaw damper coupler was replaced in response to that problem. Two days later, a pilot complained that the "yaw damper abruptly moves rudder occasionally for no apparent reason...unintended rudder input on climbout at [25,000 feet]." The main rudder PCU yaw damper transfer valve was replaced and the aircraft returned to service.

As a result of a NTSB recommendation stemming from that accident, in 1997 the National Oceanic and Atmospheric Administration (NOAA) and the National Center for Atmospheric Research (NCAR) teamed together to observe, document, and analyze potential meteorological hazards in the Colorado Springs area, with a focus on the approach paths to the Colorado Springs Airport. Mountain rotors observed during the NOAA/NCAR data gathering program had a maximum rotational rate of 0.05 radians per second, which is twelve times less than the NTSB determined would have been necessary to have produced the extreme control difficulties that brought down United 585. It was beginning to look less likely that an atmospheric disturbance caused that accident.

Eastwind Airlines flight 517

In June 1996, an Eastwind Airlines Boeing 737-200 was on approach to land at Richmond, Virginia. As the aircraft was descending through 4,000 feet msl, with an indicated airspeed of 250 knots, the aircraft yawed abruptly to the right and then rolled right. The captain immediately applied opposite rudder and "stood pretty hard on the pedal," along with simultaneously applying left aileron. The rudder seemed stiffer than usual and did not seem to respond normally throughout the event. The aircraft continued to roll right with a simultaneous heading change of about five degrees per second. The captain advanced the right thrust lever to compensate for the rolling tendency with differential power. Finally, the crew accomplished the emergency checklist, and subsequently switched off the yaw damper. The upset ended and the crew was able to safely land the aircraft.

In the month prior to this event there were three pilot-reported complaints pertaining to the rudder on this particular aircraft. Following the first event, the main rudder PCU was replaced. In response to the other pilot complaints, various components of the yaw damper were replaced, and the aircraft was test flown and returned to service. The Richmond incident happened later that night.

"The Eastwind incident became a key element of the USAir 427 investigation," stated Tomas Haueter, NTSB investigator-in-charge for the USAir 427 investigation.

Crossover airspeed

Using the same specially instrumented 737-300 that was used in the wake turbulence testing, in September 1996 a series of flight tests were conducted from Boeing Field in Seattle, Washington. Several flight conditions required test pilots to maintain control of the airplane and a constant heading by using the control wheel to oppose full rudder deflections. These tests revealed that in the Flaps 1 configuration and at certain airspeeds, the roll authority (using spoilers and ailerons) was not sufficient to completely counter the roll effects of a rudder deflected to its blowdown limit. The airspeed at which the maximum roll control (full roll authority provided by control wheel input) could no longer counter the yaw/roll effects of a rudder deflected to its blowdown limit was referred to by the test group participants as the "crossover airspeed."

The flight tests showed that when configured at Flaps 1 and weighing 110,000 pounds, the 737-300 crossover airspeed was 187 knots at one G, or normal gravitational acceleration. The significance of these weight and speed values is that at the time of the initial upset, USAir 427 weighed approximately 108,600 pounds and was operating at about 190 knots. These flight tests showed that at airspeeds above 187 knots, the roll induced by a full rudder deflection could be corrected by control wheel input. However, in the same configuration at airspeeds of 187 knots and below, the roll induced by a full rudder deflection could not be completely eliminated by full control wheel input in the opposite direction, and the airplane continued to roll in the direction of the rudder deflection. Also significant is that as G-loading increases (through increasing the vertical load factor, or angle-of-attack), the crossover airspeed increases.

The flight tests also revealed that the test airplane's rudder traveled slightly farther than originally shown by Boeing computer modeling before reaching its aerodynamic blowdown limit. Data from the flight tests were programmed into Boeing's engineering flight simulator, and numerous simulated flights were "flown" by test pilots. These simulator flights indicated that when flying at the same weight and flap configuration as flight 427, with the rudder deflected to its aerodynamic blowdown limit, the roll could not be completely stopped (and control of the aircraft could not be regained) by using full control wheel inputs at any airspeed below 187 knots.

Human performance

While the investigation of the mechanical aspects of the accident was ongoing, the NTSB's Human Performance Group was quietly conducting its own inquiry. It was felt that some of the most direct evidence of the flightcrew's actions was contained on the CVR. Unlike some investigations, the NTSB's CVR analysis wasn't all based on what the pilots said. Instead, valuable clues came from how the crew said things, as well as from other unlikely sounds on the CVR.

Speech, breathing, and other CVR-recorded sounds

To better understand the actions and emotional states of the pilots during the accident sequence, investigators extracted several acoustical signatures from .the CVR. Examined were the pilot's speech (voice fundamental frequency, amplitude, speech rate, and content) and breathing (inhaling, exhaling, and grunting) patterns that were recorded by the CVR during the routine portions of the flight, the initial upset, and the uncontrolled descent. Before this accident, it may not have been possible to use a CVR to analyze breathing rates, but in 1991 U.S. Federal Aviation Regulations mandated use of boom microphones for crewmembers on most types of aircraft. These "hot" microphones are attached to each pilot's headset and are placed directly in front of each pilot's mouth.

In addition to their own in-house human performance expertise, the NTSB sought council of three independent specialists. One specialist, Dr. Alfred Belan, a medical doctor from the Interstate Aviation Committee of Moscow, Russia, had personally participated in investigation of over 250 aviation accidents. He was recognized in the areas of medical and psychological aspects of accidents, most notably in the area of psychological analysis of speech. The other specialists were from the U.S. Naval Aerospace Medical Research Laboratory and from the NASA Ames Research Center.

After auditioning the CVR, the specialists agreed that there were no speech patterns by the pilots or other sounds to indicate that either pilot was physiologically impaired or incapacitated. Instead, the specialists agreed that both pilots sounded alert and responsive throughout the flight, including the upset and accident sequence. The specialists concluded from the captain's remark "sheeez" and the first officer's utterance of "zuh," that the pilots were surprised by the initial upset, but they further agreed that the pilots responded promptly to the situation and were attempting to control the airplane and identify the problem.

Dr. Belan noted that immediately after the initial upset began, the pilots displayed increased speech amplitude (loudness), fundamental frequency (voice pitch), frequency of breathing and reduced information within a statement. This, he stated, was an indication that their psychological stress level had increased. He explained, however, that the increased psychological stress did not necessarily interfere with the pilots' ability to respond to the emergency. In his report to the NTSB, Dr. Belan classified stress in three stages, with stage one actually improving a person's performance by "providing a constructive mobilization of attention and resources." Stage two is characterized by hasty or premature actions, such as omission of words or checklist items. However, in this stage, a person can still perform adequately, according to Dr. Belan's classification. "It is only at the highest levels of stress (third stage, or 'panic'), that the person can not think or perform clearly," Dr. Belan stated.

The NTSB noted that the captain stated the phrase "four twenty seven" during both routine and emergency transmissions to ATC, and this provided a basis for direct comparison using the same words. When the captain spoke the phrase "four twenty seven" during routine flight operations, the average fundamental frequency value was 144.6 Hz. However, when he spoke those same words during the emergency period ("four twenty seven, emergency"), the fundamental frequency had increased forty-seven percent, to 214 Hz. A forty-seven percent increase in fundamental frequency is consistent with a person's stress level being elevated into stage two, according to Dr. Belan's classification scheme. He indicated that neither of the pilot's stress level increased to stage three until about five seconds before impact, and based that finding partly on the acoustical measurements from the captain's statement, "pull...pull...pull."

Dr. Belan commented on the first officer's speech disruptions, such as grunts and forced exhalations. He reported to the NTSB that "...a person making a great physical effort develops a musculoskeletal 'fixation' (of the chest), which leads to deterioration of the normal expansion and ventilation of the lungs (inhaling and exhaling). These changes are manifested during speech. Sounds such as grunting and strain appear in speech as the person tries to minimize the outflow of air. Inhaling and exhaling become forced and rapid." He noted that normal use of cockpit controls should not produce these types of sounds, and by the first officer making these grunting sounds, it indicated he "was struggling unusually hard...[as] if he was experiencing unusual resistance in the use of a control."

Rudder pedal forces of jammed rudder systems

In June 1997, ground tests were conducted in a 737-300 to experience how rudder pedals feel when the PCU servo valve secondary slide is jammed. A special tool was used to simulate a secondary slide jam at different positions. The chairman of the NTSB's Human Performance Group for the USAir 427 accident participated and remarked, "I [applied] hard left rudder. With one or two exceptions, this input triggered a rudder reversal on the pedals. Immediately after my input, the left rudder pedal began moving outwards until it reached the upper stop...The motion was steady and continued without pause no matter how hard I pushed to counter it ('unrelenting' was a description that, at the time, seemed to capture my impression [of the force]...)."

The investigation revealed that when a jam occurs and the rudder pedal reaches its upper stop, a pilot with sufficient strength can actually move the rudder pedal towards the neutral position. This is due to cable slack and certain tolerances, known as "compliance." Because of this, the NTSB stated, "A neutral rudder pedal position is not a valid indicator that a rudder reversal in the Boeing 737 has been resolved."

Spatial disorientation

During the investigation, an exploratory hypothesis was generated that as the aircraft got tossed around by Delta 1083's wake turbulence, the USAir 427 crew may have become spatially disoriented and therefore misapplied the flight controls in their recovery attempt. To explore that issue, the NTSB conducted testing in the NASA Ames Vertical Motion Simulator (VMS). This is the world's largest motion simulator, with 60 feet of vertical motion, and 40 feet of horizontal motion. NASA Space Shuttle astronauts train in this simulator.

Several seconds of flight 427's initial upset and emergency descent were programmed into the VMS, and the CVR tape was synchronized to the motions of the simulator. A computer-generated horizon was in-place to replicate the flying conditions that existed at the time of the accident. (The NTSB interviewed the captain of Delta 1083 and learned that the in-flight visibility was unrestricted and the horizon was clearly defined). For several seconds of the upset, those investigators participating in the VMS study could actually see, hear and feel the sensations experienced by the USAir 427 crew as the upset unfolded.

The NTSB employed the use of a NASA research scientist who specialized in spatial orientation. He studied the accident history and the CVR transcripts, and then was placed in the VMS for several runs. After reviewing all pertinent data and flying through numerous simulated flights in the VMS—some with horizon visual cues present (visual display turned on) and some without these cues (visual display turned off)—the expert had formed an opinion. "I believe that the pilots probably would have experienced little difficulty in maintaining an accurate perception of their orientation, even during any brief periods when they may have lost sight of the visual horizon due to the pitch down attitude of the airplane. In addition, perturbations of the flightpath generally appear to have been followed by verbal comments from the pilots, indicating that they were fully aware of their trajectory and that they were not able to change it... [I]t does not appear at all likely that pilot disorientation due to abnormal vestibular stimulation provided a major contribution to this accident."

Conclusions

The USAir 427 flightcrew were properly certified and qualified, and were trained in accordance with applicable Federal regulations. There was no evidence of any medical or behavioral conditions that might have adversely affected the flightcrew's performance during the accident flight.

The accident airplane was equipped, maintained, and operated in accordance with applicable Federal regulations. The airplane was dispatched in accordance with FAA and industry-approved practices. All doors and hatches of the aircraft were closed at impact, and there was no evidence of an in-flight fire, bomb, explosion, or structural failure.

The aircraft encountered wake turbulence from the Delta 727 that it was following, but the wake turbulence alone could not have matched the continued heading change and flight profile that was evidenced by the FDR, computer simulation, and kinematic analysis, after 7:03:00 P.M.

The computer simulation and the kinematic analysis, which together took years of work to refine, showed that the heading changes recorded by the FDR after 7:03:00 P.M. were consistent with the rudder being deflected to its left aerodynamic blowdown limit. "Accordingly, the Safety Board concludes that, about 7:03:00, USAir flight 427's rudder deflected rapidly to the left and reached its aerodynamic blowdown limit shortly thereafter," stated the NTSB in their final report of the accident. "This movement of the airplane's rudder system could have been only been caused by a flightcrew action or a mechanical rudder system anomaly."

Rudder jam/reversal scenario

According to the FDR, between about 7:02:58 and about 7:03:00, the airplane's heading moved quickly past the assigned 100 degrees to about 94 degrees. The computer simulation and kinematic analysis indicated that this heading change was likely associated with a significant yawing motion that would have caused a lateral acceleration at the pilots' seats of more than 0.1 G to the left. The pilots would have likely felt this acceleration as a sustained, uncomfortable side force in the cockpit and would have observed the ground and sky moving sideways across the airplane's windshield.

At 7:02:59.4, the captain stated "whoa," likely in response to the kinesthetic and visual sensations produced by the airplane's yawing motion. The Safety Board concluded that it would have been reasonable for the first officer to respond to this yawing motion by applying right rudder pedal pressure about 7:03:00. "This right rudder input, intended to relieve the side force and return the airplane to its assigned heading, was instead followed by a rapid rudder deflection to the left (rudder reversal) that increased the left yawing motion and accelerated the airplane's heading change to the left," the Safety Board stated.

The NTSB concluded that as the rudder deflected to its initial blowdown position, the rudder pedals would have moved in a direction opposite to that commanded by the first officer. The first officer would likely have sensed the right rudder pedal rising underneath his right foot despite attempts to depress the pedal. During that time (between about 7:03:00 and about 7:03:02), the CVR recorded the sounds of grunting on the first officer's hot microphone channel. The NTSB tied this to the findings of the speech experts who examined these sounds and indicated that they were signs of significant physical effort. The

Safety Board recalled the words of Dr. Belan, who stated that it appeared that the first officer "was struggling unusually hard...[as] if he was experiencing unusual resistance in the use of a control." The Safety Board concluded that the soft grunting sound recorded on the CVR at 7:03:00.3, shortly after the start of the rudder reversal, was likely to be a "manifestation of an involuntary physical reaction by the first officer to the beginning of the reversing motion of the rudder pedal."

The NTSB remarked that rudder reversal, combined with the rapid left roll and yaw of the aircraft, would have undoubtedly confused and alarmed the first officer. Paradoxically, the harder the first officer pushed on the right rudder pedal, the more likely it became the jam would not clear. "A pilot pushing on the rudder pedal in this situation would know that the pedal was not responding normally, but would have difficulty comprehending, evaluating, and correcting the situation," stated the NTSB.

The NTSB stated that the information on the CVR (such as grunting) and flight control inputs used in the Safety Board's best-match computer simulation of the accident are consistent with pilot responses that might be expected during a rudder reversal.

Pilot input scenario

At about 7:02:59, just after the initial wake turbulence encounter, the rapid roll to the left was being arrested and the aircraft began to briefly roll to the right towards a wings-level attitude. The Boeing-produced kinematic analysis indicated this arresting of the left roll was in response to the first officer applying full right control wheel input. It was hypothesized that the first officer may have become startled by the resulting rapid reversal in bank from left to right, as well as the rate in which the aircraft accelerated back towards the right. To slow this rapid acceleration to the right, Boeing proposed that the pilot then applied left rudder input from about 7:02:58 to about 7:03:01.

The Safety Board countered that argument, saying, "However, such a right roll acceleration would have helped the pilots stop the left roll and regain a level bank attitude; thus there would be little reason for the pilots to have opposed the right roll acceleration." The Board pointed out that at the time of the suggested full left rudder pedal input, the right control wheel input had just stopped the airplane's left roll. "Although the right roll acceleration toward level flight might have prompted the flightcrew to remove some, or all, of the existing right control wheel input, it is unlikely that the flight-crew would have responded to this right roll acceleration by applying full left rudder before using the airplane's roll control authority to the left," concluded the Safety Board. "Moreover, both the Safety Board's computer simulation and Boeing's kinematic analysis of this period indicated that the pilots were experiencing the side load from a yaw acceleration to the left (caused by the wake vortex)." The side load would logically have prompted the pilots to apply right, rather than left, rudder. "Therefore," concluded the Board, "it [is] highly unlikely that the flightcrew of USAir flight 427 applied full left rudder, as...proposed by Boeing."

The Safety Board further concluded that "analysis of the CVR, Safety Board computer simulation, and human performance data (including operational factors)...shows that they are consistent with a rudder reversal most likely caused by a jam of the main rudder PCU servo valve secondary slide...and overtravel of the primary slide."

Unfinished business: United 585 and Eastwind 517

Since early in the USAir 427 investigation, there had been nagging concerns within the Safety Board that the USAir 427 accident, although different in some respects, had remarkable similarities with the unsolved United 585 accident. Indeed, the NTSB had learned a great deal more about investigating rudder problems in the time since they finished the 585 investigation. In addition to having more definitive methodologies for investigating potential mechanical irregularities, there was now a wealth of new human performance measures for investigation, such as speech analysis. The decision was made to quietly reopen the United 585 investigation and combine those efforts with the ongoing USAir 427 work. And then some months later, the Eastwind 517 event occurred. Many of the newly learned investigative techniques were applied to the 585 and 517 investigations. When the NTSB unveiled their findings from the USAir 427 investigation, they could show the world that they had solved those mysteries, too.

In their first statement of conclusions pertaining to the United 585 accident, the NTSB started by ruling out one of the two possible causes that they had suggested when ending that investigation: "It is very unlikely that the loss of control in the United flight 585 accident was the result of an encounter with a mountain rotor." After explaining each scenario the Safety Board stated, "The upsets of USAir flight 427, United flight 585, and Eastwind flight 517 were most likely caused by the movement of the rudder surfaces to their blowdown limits in a direction opposite to that commanded by the pilots. The rudder surfaces most likely moved as a result of jams of the secondary slides to the servo valve housings offset from their neutral position and overtravel of the primary slides." Additionally, the Board discovered that these jams could occur without leaving any physical evidence of the jam in the servo valve.

So, why did USAir 427 and United 585 crash, while Eastwind 517 was able to land safely? The answer is that Eastwind was at 250 knots when the event occurred—much above crossover speed, while the others were at or very near crossover speed (which increases as bank angle increases) for their respective configurations. When the undesired roll occurred, because they were below crossover speed, the United and USAir pilots could not counter it with roll control (control wheel input). Both aircraft rolled until they stalled. From there, the crashes were imminent.

Lack of FDR information

In accordance with then-existing Federal regulations, the aircraft involved in the United 585 and USAir 427 accidents were equipped with five- and thirteen-parameter FDRs, respectively. The NTSB was quick to point this out several times during their extensive investigation, and in February 1995, they issued an urgent recommendation (A-95-25) to the FAA to require higher-fidelity FDRs. Although the FAA acted on the recommendation, the NTSB was unimpressed with the timetable in which the FAA allowed for industry compliance. In their final of thirty-four conclusions from the 427 investigation, the NTSB addressed their displeasure: "The FAA's failure to require timely and aggressive action regarding enhanced flight data recorder recording capabilities, especially on Boeing 737 airplanes, has significantly hampered investigators in the prompt identification of potentially critical safety-of-flight conditions and in the development of recommendations to prevent future catastrophic accidents."

Probable cause

The Safety Board wrote in one paragraph a statement that took over 100,000 investigative-hours and four-and-a-half years of work, to be able to conclusively prove: "The National Transportation Safety Board determines that the probable cause of the USAir flight

427 accident was a loss of control of the airplane resulting from the movement of the rudder surface to its blowdown limit. The rudder surface most likely deflected in a direction opposite to that commanded by the pilots as a result of a jam of the main rudder power control unit servo valve secondary slide to the servo valve housing offset from its neutral position and overtravel of the primary slide."

With that, the longest aircraft accident investigation in U.S. history had ended.

Recommendations

During the course of the investigations of the United 585 and USAir 427 accidents, the NTSB issued twenty-seven recommendations to the FAA on matters dealing with the 737 rudder, pilot training, mountain rotors, and FDRs. During their final hearing on the cause of the USAir 427 accident, the NTSB issued ten additional recommendations to the FAA.

- Require that all existing and future 737s, as well as all future transport-category airplanes have a reliably redundant rudder actuation system. (A-99-20 and A-99-22)
- Convene an engineering test and evaluation board to conduct a failure analysis to identify potential failure modes, a component and subsystem test to isolate particular failure modes found during the failure analysis, and a full-scale integrated systems test of the Boeing 737 rudder actuation and control system to identify potential latent failures. The engineering board's work should be completed by March 31, 2000, and published by the FAA. (A-99-21)
- Amend the FARs to require that transport-category airplanes be shown to be capable of continued safe flight and landing after jamming of a flight control at any deflection possible, up to and including its full deflection, unless such a jam is shown to be extremely improbable. (A-99-23)
- Revise Airworthiness Directive 96-26-07 so that procedures for addressing a jammed or restricted rudder do not rely on the pilots' ability to center the rudder pedals as an indication that the rudder malfunction has been successfully resolved, and require Boeing and U.S. operators of 737s to amend their Airplane Flight Manuals and Operations Manuals accordingly. (A-99-24)

- Require all FAR Part 121 air carrier operators of the 737 to provide their flightcrews with initial and recurrent flight simulator training in the "Uncommanded Yaw or Roll" and "Jammed or Restricted Rudder" procedures in Boeing's 737 Operations Manual. The training should demonstrate the inability to control the airplane when below crossover airspeed by using the roll controls and include performance of both procedures in their entirety. (A-99-25)

- Require Boeing to update its 737 simulator package to reflect flight test data on crossover airspeed and then require all 737 operators to incorporate these changes in their simulators used for 737 pilot training. (A-99-26)

- Evaluate the 737's block maneuvering speed schedule to ensure the adequacy of airspeed margins above crossover airspeed for each flap configuration, and provide the results of the evaluation to air carrier operators of the 737 and the NTSB, and require Boeing to revise block maneuvering speeds to ensure a safe airspeed margin above crossover airspeed. (A-99-27)

- Require that all 737s operated under FAR Parts 121 or 125 to be equipped with a FDR system that records, at a minimum, the parameters required by FARs 121.344 and 125.226, dated July 17, 1997, plus the following parameters: pitch trim; trailing edge and leading edge flaps; thrust reverser position (each engine); yaw damper command; yaw damper on/off discrete; standby rudder on/off discrete; and control wheel, control column, and rudder pedal forces. Aircraft already equipped with a FDR must meet the new requirements by July 31, 2000, while those without existing FDRs have until August 1, 2001. (A-99-27 and 28)

Industry actions

These accidents placed a wake-up call to the airline industry that pilots should be better trained to recover from upsets. Although the NTSB exonerated the USAir and United crews for their performance, they also observed, "Pilots would be more likely to successfully recover from an uncommanded rudder reversal if they were provided with necessary knowledge, procedures and training to counter such an event." Today, virtually every scheduled major U.S. carrier is providing "unusual attitude" or "upset recovery" training for its pilots.

Crew procedures have been developed whereby crews should perform certain actions by memory if they experience an uncommanded roll or yaw in the 737. Additionally, many carriers have increased the Boeing-published minimum maneuver "block speeds" to provide greater margin above crossover speed.

Boeing has introduced rudder system changes to the 737. All new 737s are being delivered with the new system, and existing 737s are being retrofitted. The NTSB concluded, "When completed, the rudder system design changes to the Boeing 737 should preclude the rudder reversal failure mode that most likely occurred in the USAir flight 427 and United flight 585 accidents and the Eastwind flight 517 incident." But with that, the NTSB also issued an ominous conclusion: "Rudder design changes to Boeing 737-next-generation series airplanes and the changes currently being retrofitted on the remainder of the Boeing 737 fleet do not eliminate the possibility of other potential failure modes and malfunctions in the Boeing 737 rudder system that could lead to a loss of control."

Epilogue

In addressing the large audience that had gathered for the NTSB's final hearing of the USAir 427 accident, NTSB Chairman Jim Hall stated, "I cannot overstate the importance of our work in this investigation, because the crashes of USAir flight 427 and United Airlines 585 in Colorado Springs have raised questions in many minds about the design and operation of the 737's rudder system. The Boeing 737 is the most popular airline model in the world. There are more than three thousand of them in service worldwide, and they have amassed ninety-one million flight hours. In their thirty-one years of service, 737s have reportedly carried the equivalent of the population of the entire world—almost six billion people. At any given moment, eight-hundred 737s are in the air around the world." In fact, the Boeing 737 safety record is comparable to that of similar-type aircraft, according to a NTSB review.

It seems strange that with the 737 being in service for so many years, having flown so many miles and carried so many people, that a latent condition in the rudder system could have remained dormant for so long. But as shown by the USAir 427 and United 585 accidents, when latent conditions awaken from dormancy, it can be a very rude awakening. And deadly.

References and additional reading

Dornheim, Michael A. 1992. NTSB probes rudder anomaly as factor in United 585 crash. *Aviation Week & Space Technology*. 10 August, 29.

National Transportation Safety Board. 1999. *Aircraft accident report: Uncontrolled descent and collision with terrain. USAir flight 427. Boeing 737-300, N513US. Near Aliquippa, Pennsylvania. September 8, 1994.* NTSB/AAR-99/01. Washington, D.C.: NTSB.

National Transportation Safety Board. 1999. Public Docket. *USAir flight 427, Boeing 737-300, N513US. Near Aliquippa, Pennsylvania. September 8, 1994.* DCA94MA076. Washington, D.C.: NTSB.

North, David M. Editor-In-Chief. 1995. Actual aerobatic training a must. *Aviation Week & Space Technology*. 8 May, 66.

Phillips, Edward H. 1997. NTSB: Expedite upgrades to Boeing 737 rudder PCUs. *Aviation Week & Space Technology*. 3 March, 36.

——. 1995. Boeing 737 rudder PCUs focus of FAA directive. *Aviation Week & Space Technology*. 20 March, 33.

Wald, Matthew L. 1999. Rudder flaw cited in Boeing 737 crashes. *New York Times*. 24 March, A24.

2

Pressure falling rapidly

American 1572's brush with disaster

Operator: American Airlines

Aircraft Type: McDonnell Douglas MD-83

Location: East Granby, Connecticut

Date: November 12, 1995

As weather systems go, this one was impressive. An occluded front swept across eastern New York State and western Connecticut, spawned by a very deep low-pressure system over Quebec. A smaller low-pressure area was anchored over New York City. Strong, gusting winds and a widespread area of rain heralded the arrival of winter weather throughout New England.

Bradley International Airport (BDL), located in East Granby, Connecticut, serves the metropolitan areas of Hartford and Windsor Locks. At 9,502 feet long, Runway 6-24 is served by an Instrument Landing System (ILS) precision approach and is the runway of preference for transport-category aircraft. Runway 15-23 is only 6,846 feet in length, has an ILS available for landing to the west, and a Very high frequency Omnidirectional Range (VOR) nonprecision approach for traffic landing to the east.

At midnight on November 11, 1995, the weather at the Bradley airport was lousy. Skies were overcast, with five miles visibility, light rain, and winds out of the southeast at 28 knots gusting to 42 knots. The altimeter setting was 29.42 inches Hg. and the barometric pressure was falling rapidly. Low-level wind shear advisories (LLWAs) were in effect, and four notices (SIGMETs, or significant meteorological information, and AIRMETs, airman's meteorological information) were issued for severe turbulence and icing at lower altitudes. Conditions were deteriorating. Fortunately, almost all of the flights scheduled to arrive at Bradley that evening were parked safely at their gates. All but one.

Flight history and background

American Airlines flight 1572 was late. The Washington, D.C., based flightcrew was on the second day of a three-day trip and were almost two hours late leaving Chicago for Bradley Airport.

Prior to departure, the pilots had been briefed regarding the marginal weather conditions all along the East Coast. As part of a normal flight plan review, their dispatcher had sent via the automatic communication and recording system (ACARS) the latest weather report at BDL. That report included the advisory "pressure falling rapidly," which indicated a barometric pressure drop of at least .06 inch Hg. per hour. Typically, high winds, rain, deteriorating visibility, and rapidly changing weather conditions are associated with this meteorological phenomenon.

While en route, the crew received two additional dispatcher-generated ACARS messages regarding conditions at BDL. The first relayed required altimeter settings to be used by the crew, and the second was an advisory that airplanes had been landing, but only after experiencing wind shear and turbulence on approach. The first altimeter setting sent to the crew indicated a local pressure of 29.42 inches Hg., which would be used to indicate altitude above mean sea level (msl) and is referred to as "QNH." The second setting was a "converted" pressure of 29.23 inches Hg., which the crew would use on a different altimeter to show "above field elevation" measurements, or "QFE."

American Airlines' MD-83s (see Fig. 2-1) were equipped with three altimeters; the airline's policy was that for all departures and arrival operations, both pilots' primary altimeters were to be set to QFE, thus showing altitude above the field elevation. The standby altimeter was always set to QNH, showing aircraft altitude above mean sea level. Typically, all altimeters in an airliner cockpit would be set to show mean sea level at all times; American was the only U.S. airline to use this particular altimeter setting procedure. The two settings issued, then, if correct, would cause both primary altimeters to indicate "0" feet with the aircraft on the ground at BDL (or almost any airport, for

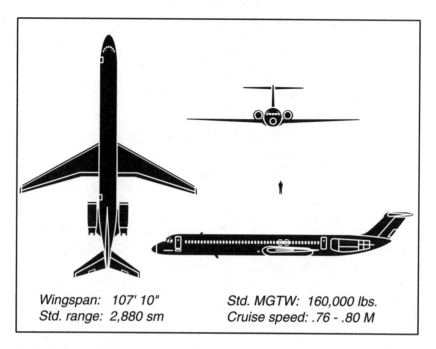

Wingspan: 107' 10"   Std. MGTW: 160,000 lbs.
Std. range: 2,880 sm   Cruise speed: .76 - .80 M

2-1 *McDonnell Douglas MD-83.*

that matter), and "174" feet (the height above mean sea level at Bradley Airport) to be displayed on the standby altimeter. Because of mechanical limitations within the instruments, however, there were a few higher altitude airports where this procedure cannot be used.

A short time after the crew received the dispatch message, the air route traffic control center (ARTCC) cleared flight 1572 to descend to 19,000 feet (FL190). The most recent BDL weather report, received by the flightcrew via the recorded automatic terminal information service (ATIS) indicated a local altimeter of 29.50 inches Hg. (QNH) and an advisory for severe turbulence. The crew noted that this information was about an hour-and-a-half old and as such might be outdated. They were then cleared to 11,000 feet msl and given the latest altimeter setting of 29.40 inches Hg.

Seven minutes later, the flight attendants were advised of possible turbulence and the "before landing" checklist was read.

When the challenge "altimeters?" was made, the captain responded "twenty nine fifty." The first officer then reminded, "they called twenty nine forty seven when we started down...whatever you want."

"OK" was the captain's only reply.

A briefing was then conducted regarding the planned VOR approach. The captain, in reviewing the chart, said "one seventy four's the elevation, so twenty nine, twenty three. Set and cross checked." The first officer replied, "minus, uh," followed immediately by the captain's comment "showing seventy...check seventy feet difference." The flight was handed off to approach control, and was cleared to 4,000 feet. Winds were reported as "one seven zero at two nine [knots] gusts to three nine [knots]," and a vector was given to intercept the final approach course.

Very strong winds had caused flexing of a control tower window, allowing rainwater into the control room. Although the resulting electrical hazards necessitated closing the facility, one supervisor did remain in the tower cab to monitor repairs. The flightcrew of 1572 was advised of the closure and was cleared for the VOR approach. The time was now ten minutes before one o'clock in the morning.

The captain was flying the aircraft using the autopilot, and wing flaps had been selected to the eleven-degree position shortly after intercepting the 328-degree radial of the Bradley VOR. While flying inbound, however, the VOR track mode of the autopilot was unable to maintain the proper course due to the very strong cross winds. Realizing this, the captain then changed the mode to "heading select" in order to manually steer the aircraft using the heading "bug" (see aircraft track representation in Fig. 2-2). Once past MISTR, the initial approach fix, the landing gear was lowered and the flaps were selected to their normal landing configuration of forty degrees. A descent was initiated using the vertical speed mode of the autopilot, at a rate of 2,000 feet per minute. By then, the aircraft was being bounced around quite a bit, having entered the area of moderate turbulence, and heavy rain was pelting the cockpit windows.

At the charted altitude of 2,000 feet msl the airplane leveled off and continued inbound towards the airport. At DILLN, the final approach fix, the crew could descend to the minimum descent altitude (MDA) published for the approach. In this case, that minimum was 1,080 feet msl, or 908 feet above the runway elevation of 172 feet.

Upon reaching DILLN, the captain called for "a thousand down" in the vertical speed display, which pitched the aircraft over slightly and resulted in a 1,000 feet per minute descent. The nose gear landing lights were turned on briefly, but the bright reflection back into

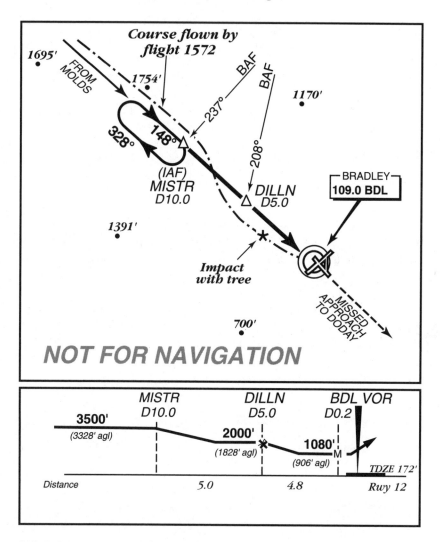

2-2 *VOR Runway 15 approach at Bradley and American 1572's flight path....*Source: NTSB

the cockpit further decreased forward visibility, so they were immediately extinguished. The tower supervisor issued a wind shear alert for the airport, and at about the same time the first officer called "there's a thousand feet, you got forty forty land, cleared to land." This was to alert the captain that the aircraft was properly configured for landing, and that he was only 1,000 feet above the field elevation. Five seconds later the first officer stated "...now nine hundred and eight is your uh..." to which the captain replied, "right."

The first officer then began looking outside to try to see the approach lights extending from the end of the runway. Looking straight down, he was able to see the ground, but the bases of the clouds obscured his view forward. The airplane was continually jolted by the turbulence. Glancing back inside the cockpit to the flight instruments, the first officer noticed that the aircraft was beginning to descend below the minimum descent altitude of 908 feet (above runway elevation). He quickly said, "you're going below your..." The captain then reached up and selected "altitude hold" on the flight guidance panel, using the autopilot to maintain the current altitude (see Fig. 2-3).

Just then, the ground proximity warning system (GPWS) detected an excessive rate of descent for the airplane's configuration and altitude, and initiated it's characteristic warning. With "sink rate, sink rate!" continually blaring through the cockpit speakers, the captain pulled on the control column and simultaneously moved the throttles forward to the EPR (engine pressure ratio) limit. The aircraft slowly started to respond to control inputs, but four seconds later flight 1572 flew through the tops of oak trees on Peak Mountain Ridge. The first tree was struck seventy-six feet above the ground, at an elevation of 819 feet msl and about two and a half miles from Runway 15.

The pilots felt a jolt and heard a loud report, but a "wind shear, wind shear" alarm drowned out most other noise. The captain immediately started to "go around," shoving the throttles all the way forward to the mechanical stops while calling for the flaps to be raised to fifteen degrees and the landing gear to be retracted. The wind shear alert continued undiminished for another few seconds, and both a horn and a synthesized voice warned "landing gear, landing gear!" as the landing gear selector was placed to the "up" position.

They were still flying, but damage to the aircraft was significant. Fuselage fairings were punctured and gouged, and branches had severely torn and dented the leading edges of the wings. Both engines had ingested tree limbs, causing catastrophic internal damage within the left engine and a fire inside the compressor section of the right engine. The flaps were extensively damaged and the right main landing gear door was ripped off.

Twenty seconds later, the captain called out "the left motor's failed!" Both engines were experiencing severe compressor stalls, and the disturbed airflow produced deafening banging and popping noises.

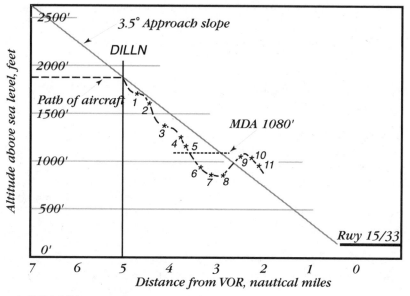

1. CAM-2 "You got a long way to go"

2. TWR   "Wind shear alert..."
   RDO-2 "Copy"

3. CAM-2 "There's a thousand feet. You got forty forty land, cleared to land."
   CAM-1 "OK."

4. CAM-2 "Now. Nine hundred and eight is your, uh..."
   CAM-1 "Right."

5. CAM-2 "Your * bug"
   CAM   [Sound of rattling similar to aircraft going through turbulence]

6. CAM-2 "You're going below your..."
   CAM-4 "Sink rate...sink rate"

7. CAM   Sound of impact
   CAM-4 "Wind shear, wind shear..."

8. CAM-2 "Go...go around!"
   CAM-1 "We're going, going, going around, going around!"
   CAM-4 "Landing gear..."

9. CAM-1 "Left motor's failed!"

10. CAM-2 "There's the runway straight ahead."
   CAM-1 "OK."

11. CAM-1 "Tell 'em we're goin' down. Tell 'em emergency."
   RDO-2 "Tower call for emergency equipment. We have, we're goin down
           on the runway!"

CAM Cockpit Area Microphone	**RDO - 2** F/O Radio Transmission	
CAM - 1 Captain	**TWR** Radio Transmission from	
CAM - 2 First Officer	Control Tower	

2-3 *Profile view of American 1572's approach path correlated to CVR*

...Source: NTSB

The right engine began to lose power. To maintain airspeed, a shallow descent was started. But as the rain abated and the airplane settled, the first officer saw the airport. "There's the runway straight ahead!" he yelled.

Struggling to maintain control, the captain said "tell 'em we're going down, tell 'em emergency!"

"Tower, call for emergency equipment. We have...we're goin' down on the runway!" the first officer called. He then asked the captain if the landing gear should be lowered again.

"Yes, throw it down!" he confirmed. The GPWS, meanwhile, had started to shriek out a new "sink rate, sink rate" warning over the cockpit speakers.

"You're gonna make it!" offered the first officer, "keep coming..." He then asked "flaps?" "Put 'em down!" was the response.

The GPWS screamed out yet a new warning, "Too low, flaps, terrain, terrain...too low! Don't sink!"

"Flaps, flaps forty. All the way down," yelled the captain. He hoped that full flaps would provide a "balloon effect," extending his glide distance to a point closer to the airport.

"They're all the way."

"OK, hold on guy!" shouted the captain just a moment before the airplane clipped a small tree off the end of the runway. A shower of flame and sparks flew out of the right engine, and the pilots fought to align their crippled aircraft with the pavement.

The plane slammed into a row of metal posts and supporting concrete pads for the ILS localizer antenna, bounced through the overrun area and up onto the paved surface. One tire burst and hydraulic lines were torn from the landing gear struts. The rear of the aircraft struck the ground, crushing the lower part of the fuselage and damaging the aft pressure bulkhead. Initially tracking to the left of the centerline, the aircraft then veered back toward the middle of the runway.

"Get it on, on the deck!" the captain cried out. "Hold it down buddy, hold it down, hold it down, hold it down..." As they began to slow down, the fire handles for both engines were pulled, shutting off all fuel. The aircraft finally stopped about 3,500 feet from where it had first touched down.

The captain then immediately turned on the Emergency Power switch, picked up the public address system handset and ordered "easy victor, easy victor," commanding the cabin crew to begin an evacuation. He also advised passengers not to use the aft stairs and called for the evacuation checklist.

As the evacuation started, one of the flight attendants instructed passengers to remove their shoes. Most complied and rapid egress from the airplane was soon impeded by the stacks of shoes in the aisles and near the emergency exits. The slide at the aft emergency exit/galley door did not inflate automatically when the door was opened, but did inflate when the manual inflation handle was pulled by the flight attendant.

The first officer finished his duties in the cockpit prior to the captain, and exited out the forward door and down the slide. The captain then left the flight deck and proceeded through the cabin, checking the lavatories to assure that no passengers remained in the aircraft. He also evacuated via the forward passenger door.

The investigation and findings

For one week after the accident, investigative teams specializing in meteorology, air traffic control, aircraft performance, structures, systems, operations, and powerplants worked at the scene under the direction of the National Transportation Safety Board (NTSB). Additionally, cockpit voice and flight data recorder groups met in Washington, D.C.

Areas of particular interest to the accident investigators included the following:

Adequacy of weather information

Several current SIGMETs had not been passed along to the crew, but equivalent information was included in the initial American Airlines flight dispatch document. The weather at BDL upon arrival was as forecast, with strong winds, moderate turbulence, and possible wind shear. The crew also received an ACARS weather update as well as the current ATIS report for the airport. That ATIS information was over an hour-and-a-half old, however, so the Safety Board believed it would have been prudent for the flightcrew to have requested the latest weather report once in contact with the BDL approach controller.

Altimeter settings

The altimeter settings of 29.42 inches Hg. (QNH) and 29.93 inches Hg. (QFE) received from the dispatcher were an hour old by the time the crew started their descent. At the time of the accident, the setting from the Boston air traffic control center (29.40 inches Hg.) was twenty-two minutes old, and the ATIS setting (29.50 inches Hg.) had been recorded almost two hours previously. Because incorrect altimeter settings were used, the Safety Board determined that the aircraft was actually seventy-six feet lower at the MDA than was being displayed on the primary altimeters.

Approach controllers are always required by the Air Traffic Control Handbook to issue a current altimeter setting when initial contact is made with an aircraft, but in this case, it was not done. As with the rest of the weather information, the NTSB determined that the crew could and should have requested this information anyway.

Descent below MDA

At certain points during any approach the pilot not flying is required to "callout" to the flying pilot the airplane's altitude and position. In this case, these callouts were required at one thousand feet above field level, one hundred feet above the MDA, and finally, the MDA itself. The first officer made the first callout when he said, "and there's a thousand feet." He was also aware of the rapidly approaching MDA when he said, "now nine hundred eight is your, uh..." and "you're going below your..." Investigators believed that if the first officer had more carefully monitored the approach on instruments and delayed looking outside for the runway environment until the airplane was level at the MDA, he would have been better able to notice the descent below MDA and call the captain's attention to it. Ultimately, the NTSB determined that the flightcrew allowed the airplane to descend 309 feet below the indicated MDA.

A Visual Descent Point (VDP) provides the crew a specific point in space during an approach at which to descend from the MDA, using a normal flightpath angle, for landing on the intended runway. Planning an approach with a VDP allows a reduced rate of descent, since the crew knows an exact distance at which they can expect to leave the MDA. This slower descent provides for easier flightpath monitoring and a more precise aircraft level-off at the MDA. VDPs can easily be calculated by the crew, and instructions for doing so are contained in all American Airlines' flight manuals. Additionally, VDPs are frequently designed into nonprecision approach procedures and are then depicted directly on the chart. But no VDP was illustrated for the BDL VOR Runway 15 approach, and the crew did not determine one on its own.

In simulator studies, the Safety Board discovered that when the autopilot is engaged with the "rough air" feature activated, automatic level off using the "altitude hold" selector can result in a continued (though temporary) descent. With a 1,000-foot rate of descent commanded, engaging altitude hold caused the simulator to descend an additional 120 to 130 feet before climbing back to the requested altitude. Taking control manually, however, did result in an immediate level-off of the simulated flightpath.

The Board also investigated the possibility that wind shear had caused deviation below the MDA. Analysis of the pressure altitude trace of the FDR indicated only a minor decreasing headwind shear that would not have significantly affected the airplane's flightpath.

Terminal Instrument Approach Procedures (TERPS)

Within the Federal Aviation Administration (FAA), the Office of Aviation System Standards is responsible for the design of instrument approaches. These specialists do so with the aid of charts, graphs, and local area maps, but not by actual physical surveys. During the initial development of the BDL VOR 15 approach, it was planned that a VDP would be located at the point where the visual approach slope indicator's (VASI) glidepath intercepted the MDA. But using local terrain charts, an obstruction was found that penetrated the "obstacle clearance plane" by fifty-five feet. So the FAA could not include the VDP in the final version of the published approach.

The Flight Inspections Operations Division of the FAA is responsible to confirm that any obstacle marked on the charts is at the height and position indicated, and that all descents and turns required by the approach procedure are within accepted operational limits. All of this is done by physical survey and flight testing. When the Atlantic City, N.J., Flight Inspection Area Office initially verified the then-new VOR Runway 15 procedure, and then again after the accident, the entire approach was found to be "obstacle free."

It is noteworthy that one FAA department found an obstruction but another division within the same organization did not. Unfortunately, the two departments never compared notes. The Safety Board pointed out that if, in fact, the obstruction existed, then the FAA's own policies should have forced a re-examination of the procedure to assure adequate clearances, and the VASI should have been relocated or decommissioned. If, however, there was no obstruction, then a suitable VDP could have been published for the approach. In any case, the NTSB found that there was a serious lack of coordination between the FAA's procedures development program and their flight inspection program.

The TERPS handbook states that anytime the interaction of strong winds and precipitous terrain might cause altimeter errors or pilot control problems, modifications to the approach procedure must be considered. Since Runway 15 was used primarily when strong winds precluded the use of Runway 6-24, the Safety Board was of the opinion that the FAA should have taken into account Peak Mountain ridge to the west of the airport and its effect on a pilot's ability to fly the approach.

The Board also found that the depiction of only one 819-foot obstacle on the published approach chart is misleading; the entire ridgeline is an obstacle. Therefore, the NTSB reiterated its support for the graphical presentation of terrain information on all approach and navigation charts. (Because of the American Airlines accident near Cali, Colombia, the NTSB published a recommendation regarding this topic several months prior to the Bradley report being finalized. See Chapter 3).

Airport issues

The ATIS report that was received by flight 1572 at 12:32 A.M. was of little use to them. It had been recorded at 10:51 P.M., almost two hours prior to the accident. It had not been updated because conditions inside the tower made communication difficult, if not impossible. However, the Safety Board was concerned that even though it was not a factor in this accident, tower controllers should make every effort to record a new ATIS anytime it is necessary, including one to indicate that the tower is about to close.

The Board also determined that one sensor of the low-level wind shear advisory system (LLWAS) was physically out of alignment, perhaps degrading the ability of the system to detect all shears. It was not deemed to be a contributing factor, however, because wind shear alerts had been issued using the remaining functional sensors all evening, and additional alerts would probably not have caused this flightcrew to abandon the approach. A malfunctioning or misaligned sensor can take up to six months to be repaired, and in the meantime, the FAA was allowed to recertify as "operational" the overall LLWAS. The NTSB was concerned that the FAA's recertification process did not address the operational and functional capabilities of each component of the LLWAS system.

Flight after the tree strike

The Safety Board was impressed with the crew resource management (CRM) and flying skills of the crew in their handling of the aircraft after impact with the trees. As they stated in their report, "Any mistakes in glidepath management, ground path management or airplane configuration timing by the captain and the first officer would probably have caused the airplane to land in unsuitable terrain...and would have resulted in more severe injuries to crew and passengers."

Evacuation issues

The escape slide system is designed to operate automatically anytime the cabin door is opened and the system is armed. As the door rotates outward, a girt bar anchors the slide package to the floor, and the physical action of the slide falling out of the aircraft pulls an inflation cable, releasing compressed air into the side and inflating it. To operate properly, the cable must pass through a grommet located on a tab near the girt bar, then be securely attached to the manual inflation handle. On the accident aircraft, the inflation cable did not pass through the grommet, allowing the slide package to fall without causing enough tension on the cable to activate the inflation mechanism. The Safety Board found that the slide system installation instructions as published in the Douglas Maintenance Manual were ambiguous and could easily be misunderstood.

During an emergency evacuation, it is imperative that all passengers move as rapidly as possible to the exits, jump out the exit and down the slide quickly, and move away from the aircraft. In the past, high-heeled shoes have punctured slides and injured rescue personnel assisting at the bottom of the slide. But modern evacuation slide design and strengthened fabrics have made this policy out-of-date, and the wearing of most types of shoes allows rapid movement away from the aircraft.

Because the discarded shoes did hinder easy access to the aisles and the exits in this evacuation, the Safety Board is concerned that there is no uniform "shoe-removal" policy among the airlines, nor has the FAA provided operators adequate guidance on this subject.

Conclusions and probable cause

The NTSB concluded that the flightcrew was qualified, the aircraft was properly maintained, the crew received accurate and appropriate weather information, the weather was above landing minimums at BDL, and there were no malfunctions of the autopilot or flight control systems that contributed to the accident. Other findings included the fact that the various altimeter-setting errors did not affect the ultimate sequence of events, and neither the winds nor wind shear were severe enough to cause deviation below the MDA.

As stated in the final report, "The National Transportation Safety Board determine[s] that the probable cause of this accident was the flightcrew's failure to maintain the required minimum descent altitude until the required visual references identifiable with the runway were in sight. Contributing factors were the failure of the BDL approach controller to furnish the flightcrew with a current altimeter setting, and the flightcrew's failure to ask for a more current setting."

Recommendations

The NTSB published fifteen separate recommendations as a result of the investigation. Two were issued in June of 1996:

- Publish a VDP for the BDL VOR Runway 15 approach, or if not possible due to obstacles, warn all pilots about the 3.5-degree glidepath angle and the high terrain along the approach path. (A 96-31 and 32)

The others were issued in conjunction with the final accident report, and included:

- Evaluate the possible use of constant rate or constant angle descents in all nonprecision approaches, in lieu of traditional "step-down" descent procedures. (A-96-128)

- Improve coordination between the flight inspection and procedures development departments within the FAA. (A-96-129)
- Better define the guidelines concerning precipitous terrain criteria and adjustments within TERPS, and to incorporate them in the BDL VOR 15 approach. (A-96-130 and 131)
- Review the appropriateness of letdown altitudes of any approach near high terrain, and solicit comments from the users of the system for all approaches, to better evaluate actual operational design. (A-96-132 and 133)
- Require controllers to issue frequent altimeter setting changes anytime the official weather report includes "pressure falling rapidly." (A-96-134)
- If time permits, require tower controllers to record a new ATIS indicating that the tower is closed. (A-96-135)
- Develop a procedure for timely repair of LLWAS systems and ensure that the recertification process addresses functional capability of all components of the system. (A-96-136)
- Require inspection of all MD-80 and DC-9 aircraft for proper rigging of the evacuation slides, and to rewrite maintenance manuals to provide clear and consistent instructions. (A-96-138 and 139)
- Develop a uniform policy on shoe removal during passenger evacuations and train flight attendants in that policy. (A-96-140)

Industry action

The FAA issued Change 17 to the U.S. Standard Terminal Instrument Procedures (TERPS) to require that descent angles and gradients be computed for all nonprecision approaches. Subsequently, depiction of the optimum three-degree descent gradient on government-issued approach charts will allow pilots to fly a rate of descent that ensures a stabilized final approach path, thus eliminating most step-down fixes. These changes will be made to nonprecision approaches during their regular biennial reviews.

Additionally, criteria have been developed for design of stand-alone area navigation (VNAV) instrument approaches using barometric vertical navigation and guidance, with a vertical path angle for each approach specified for use all the way down to the runway threshold. A "decision altitude" will replace the traditional MDA on this type of approach, because allowances are made for the loss of altitude during a missed approach, and appropriate obstacle clearance assessments have been made.

The BDL VOR Runway 15 approach has been changed to incorporate all existing precipitous terrain adjustments as found in the TERPS handbook. Additionally, the Government Industry Charting Forum, in conjunction with the FAA, is developing a plan to completely revise these guidelines. The National Center for Atmospheric Research will submit a report detailing a proposed plan of action in 1999. A VDP for the BDL 15 VOR approach was published at 2.3 miles from the end of runway 15 at BDL, coinciding with the intersection of MDA and the installed 3.5-degree VASI system for the runway.

The FAA now requires controllers, through Order 7110.65L "Air Traffic Control," to issue any changes to the altimeter setting to aircraft executing nonprecision approaches when the weather report contains the statement "pressure falling rapidly." Additionally, the FAA has revised the LLWAS maintenance procedures found in Order 6560.13B, "Maintenance of Aviation Meteorological Systems and Miscellaneous Aids." The new policy requires a complete check of the system every thirty days, reduced from ninety days, and notification of any corrective action taken to repair components of the system must also be made within thirty days.

Flight Standards Information Bulletin 97-11, "Evacuation Slide Rigging Procedures," was issued by the FAA to direct inspectors to verify that all MD-80 and DC-9 aircraft slides are properly rigged. McDonnell Douglas also agreed to review all maintenance manuals for clear, concise, and consistent slide system graphics and terminology. Another notice, Flight Standards Bulletin 97-07, was issued to all FAA Principal Operations Inspectors (POIs) for distribution to all air carriers, and states that shoes should be left on during emergency evacuations. With enough warning, high-heeled shoes should be removed and placed in an overhead bin, not in a seat-back pocket.

Actions taken by American Airlines

Just thirty-eight days after the BDL accident, American Airlines flight 965, a B-757 en route to Cali, Colombia, hit a mountain during a nighttime descent. All but four of the 163 passengers and crew were killed (see Chapter 3).

A controlled flight into terrain (CFIT) accident is one in which a mechanically normally functioning aircraft is inadvertently flown into the ground, water, or an obstacle. Because of these two CFIT accidents, American Airlines took immediate action. A bulletin entitled "Non-Precision Approach Crew Coordination Procedures" was issued to all MD-80 pilots, which reiterated the proper procedures of leveling off above rather than below the MDA, and of carefully monitoring aircraft altitude at all times during the approach. Pilots were also reminded that the pilot not flying should be focusing attention outside the aircraft only after leveling-off at the MDA.

Three temporary operational changes also were implemented:

1. Increasing nonprecision approach MDAs and visibility requirements by one hundred feet and one-half mile.

2. Increasing visibility minima on all nondirectional beacon (NDB) nonprecision approaches by one mile.

3. Raising the ceiling of the "sterile cockpit," or that time during which there can be no distractions to the pilots on the flight deck, from 10,000 feet to 25,000 feet in Latin American airspace.

The airline also established a Safety Assessment Program that would carefully examine seven specific areas of American's operations. Safety assessment teams focused on:

- Human factors issues, especially those related to approach charts, procedures, and training.

- Geographical issues relevant to each of the airline's areas of operations.

- FMS and GPWS advanced technology.

- Corporate culture and its influence on flight operations.

- Integration of Flight Operations Quality Assurance (FOQA) and Airline Safety Action Partnership (ASAP) programs into Advanced Qualification Program (AQP) training and confidential reporting systems in general.

- Improvement of communication between the various levels of flight operations.

Quarterly progress meetings were conducted, and a final report was issued in 1997. A CFIT Task Force was established by American Airlines, and specific guidance was developed to improve CFIT scenarios during simulator training.

Internal debate also resulted in American Airlines modifying its altimeter setting procedure to conform to the rest of the industry. QNH is now the accepted primary altimeter setting to be used for all phases of flight operations.

Epilogue

Flight 1572 was lucky. The aircraft was returned to service after nine million dollars worth of repairs, and no one on board was seriously injured. But this is only one of the several CFIT accidents that will be reviewed in this book and only one of hundreds that have occurred over the years. Most are much more tragic. CFIT continues to be the leading cause of transport-category aircraft accidents, and fatalities, worldwide.

References and additional reading

Federal Aviation Administration. 1999. *NTSB Recommendations to FAA and FAA Responses Report.* <http://nasdac.faa.gov/>

Flight Safety Foundation. 1997. During nonprecision approach at night, MD-83 descends below minimum descent altitude and contacts trees, resulting in engine flameout and touchdown short of runway. *Accident Prevention.* April, 1-15.

Garrison, Peter. 1997. Tree-topping. *Flying.* June, 93-94.

Hughes, David. 1996. MD-83 Tree strike sparks NTSB probe. *Aviation Week & Space Technology.* 1 January, 44.

National Transportation Safety Board. 1996. *Aircraft Accident Report: Collision with Trees on Final Approach. American Airlines Flight 1572, McDonnell Douglas MD-83, N566AA, East Granby, Connecticut, November 12, 1995. NTSB/AAR-96/05.* Washington, D.C.: NTSB.

National Transportation Safety Board. 1996. *Public Docket. American Airlines Flight 1572, McDonnell Douglas MD-83, N56AA, East Granby, Connecticut, November 12, 1995. DCA96MA008.* Washington, D.C.: NTSB.

Phillips, Edward H. 1996. Descent below MDA tied to MD-83 accident. *Aviation Week & Space Technology.* 18 November, 36.

3

El Deluvio claims American 965

Operator: American Airlines
Aircraft Type: Boeing 757-233
Location: Near Cali, Columbia
Date: December 20, 1995

Soaring over twenty-thousand feet into crystalline blue skies, the Andes Mountains traverse the entire length of South America, hugging the Pacific coastline from Cape Horn to Panama. This majestic range is the longest in the world and totally dominates the rugged landscape of western Colombia, which lies at the northwest corner of the continent.

The city of Cali is the economic, cultural, and political center of the Valle del Cauca region and is nestled at the end of a long river valley. Alfonso Bonilla Aragon International Airport serves the city; its one main north-south runway has landing approaches that require careful navigation and precise descents between high mountain ranges. Thirty-five miles to the north, the peaks of El Deluvio and San Jose stand as steadfast sentries at the narrow entrance of the Cauca river basin.

Since the pioneering flights of Antoine de Saint-Exupery, these mountains have been notoriously unforgiving of any carelessness or neglect, claiming the lives of thousands of airmen and their passengers. And on December 20, 1995, as the last rays of the setting sun flashed off those snow-covered peaks, no one could have imagined the horrible price that would again have to be paid.

Flight history and background

It was only five days until Christmas, and the two pilots arrived at the Miami International Airport dispatch office in high spirits.

Days off had been spent in typical fashion, mostly enjoying the holidays with family and friends. They had arrived forty minutes earlier than their required check-in time, and their banter reflected their friendship and anticipation of a pleasant flight to Cali. American Airlines flight 965 was scheduled to push back from the gate at 4:40 P.M. (EST).

The captain was well liked by colleagues and was thought of as a very competent and professional pilot. His file contained numerous letters complimenting his job performance and communications skills. He enjoyed international trips and had previously flown into Cali thirteen times.

While still in the Air Force, the first officer served as a flight instructor on many types of high-performance aircraft, with his abilities earning him the distinction of "Air Force Instructor of the Year" in 1985. Together the crew had thirty-five years of experience flying for American, and almost 19,000 total hours at the controls.

The crew checked the en route and destination weather and completed their flight planning. Indeed, it would be a beautiful evening to fly, and the forecast for Cali called for clear skies and mild temperatures. The crew then boarded the aircraft.

The Boeing 757-233 (see Fig. 3-1) to be used that Wednesday afternoon was very new by aviation standards. Delivered to the airline on August 21, 1991, N651AA was just over four years old and had accumulated only 13,782 flight hours on its airframe. The B-757 aircraft had a reputation for safety and reliability—at that time no 757 had ever been involved in a fatal accident. The previous flight brought the airplane in from Guayaquil, Equador, and there were no items noted in the logbook that would require maintenance before this afternoon's departure.

Aircraft number 651, like all 757s and 767s, was equipped with a state-of-the-art flight management system (FMS). This system is made up of a flight management computer (FMC), moving map displays, and other flight instruments and controls, and includes an on-board worldwide database of relevant navigational aids and airports. The FMC combines the stored information, flight plan data input from the flightcrew, and position information from the on-board computers to optimize the flight profile.

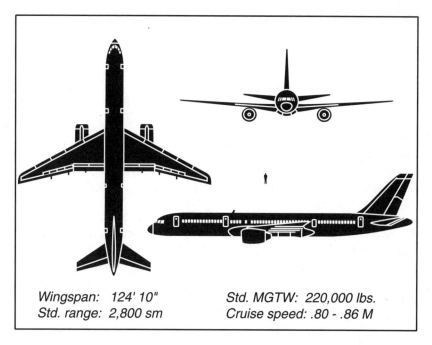

Wingspan: 124' 10"   Std. MGTW: 220,000 lbs.
Std. range: 2,800 sm   Cruise speed: .80 - .86 M

3-1 *Boeing 757-200.*

Each pilot station also has a control display unit (CDU) with a small keyboard for inputs to the computer. An integral cathode ray tube displays a vast amount of data, including aircraft position, planned route, airports, waypoints, selected navigational aids, and pilot inputs. When directed by the flightcrew, the FMC can be engaged to control all autothrottle and autopilot functions.

Once their flight kits had been stowed, the pilots began the familiar ritual of cockpit preflight checks. After programming the planned routing into the FMC, briefing the cabin crew, and completing the checklists, the flight was ready to depart on time. However, connecting passengers and baggage delayed their gate departure by thirty-seven minutes, and once they began taxiing towards the departure runway, airport ground congestion postponed their takeoff by another hour and a half. Finally airborne at 6:35 in the afternoon, the planned three-hour-and-twelve-minute flight time meant that the flight 965 would be very late arriving. This would be an inconvenience for the passengers, but it had serious operational implications, as well. Because the flight attendants were scheduled to receive only a minimum layover in Cali that night, any further delay in their arrival time could mean severe disruptions in flight schedules the following day. The pressure was on to make up as much time as possible en route.

The flight proceeded uneventfully south from Miami, through airspace controlled by Cuba, Jamaica, and finally into Colombia. Shortly before descending out the assigned cruising altitude of Flight Level 370 (FL370 or 37,000 feet above mean sea level, msl) the crew checked the current weather conditions in Cali via the onboard Aircraft Communications Addressing and Reporting System (ACARS). It still looked great—scattered clouds and visibility greater that ten kilometers.

The Cali airport had two instrument approaches that can be used for landing. One was a VOR DME (Very high frequency Omnidirectional Range, with Distance Measuring Equipment) nonprecision approach to Runway 19, for landings to the south. The other was an ILS (Instrument Landing System) precision approach to Runway 01, which was normally used by transport aircraft. The localizer portion of the ILS allowed very accurate alignment with the runway, and the glide slope provided the flightcrew information that permits very accurate descents. But for aircraft arriving from the north like flight 965, utilizing Runway 01 would require flying south of the airport for several miles before turning inbound to the runway, adding a few extra minutes to the flight time.

The first officer was the pilot flying American 965 that evening, so the captain handled all radio communications.

"Buenos noches senor, American nine six five leaving [flight level] two three zero, descending to [flight level] two zero zero. Go ahead sir," he advised Cali Approach Control.

The approach controller, located in a small room in the airport control tower, asked, "distance DME from Cali?" to which the captain replied "The DME is six three [nautical miles]." Normally, aircraft would be displayed on the controller's radar screen, making it unnecessary for the pilot to report distance from the station. But no air traffic control radar was available that night, as antigovernment guerrillas had destroyed the only operable radar facility in 1992.

"Roger, cleared to Cali VOR, descend and maintain one five thousand feet, altimeter three zero two two...no delay expect for approach. Report Tulua VOR."

The captain responded, "OK, understood. Cleared direct to Cali VOR. Report Tulua and altitude, that's fifteen thousand, three zero zero two. Is that all correct sir?"

"Affirmative" was the only reply. The local time was 9:35 P.M. (EST).

The crew then programmed the Cali VOR into the aircraft FMS as an active navigational waypoint. With only a few keystrokes, the onboard computers changed the route displayed to the pilots, showing a straight line from the present position direct to the Cali VOR station, located only eight miles south of the airport. The captain informed the first officer that the change has been made, and this input directed the aircraft, via the computer, to fly the new route.

About a minute later, the controller asked "Sir, the wind is calm. Are you able to approach runway one niner?" Because the winds were calm, the controller could offer the crew either of the runways. He recognized a landing to the south could save a substantial amount of time for the American flight. After a brief discussion, the pilots decided to accept the offer to land "straight-in" to Runway 19. At that time, the first officer stated to the captain that they would need to get lower altitude, and the captain replied to the controller's query, "Yes sir, but we'll need a lower altitude right away though."

"Roger. American nine six five is cleared to VOR DME approach runway one niner. Rozo number one arrival. Report Tulua VOR."

Flight 965 acknowledged, and the pilots started to prepare for the approach. The controller asked for a verification of the next reporting point, and the captain responded, "report Tulua."

The captain commented to the first officer, "I got to give you to Tulua first of all. You, you want to go right to Cal...er, to Tulua?"

The first officer responded, "Uh, I thought he said the Rozo one arrival?"

"Yeah, he did. We have time to pull that out?"

The cockpit voice recorder (CVR) then recorded the sound of rustling pages as both pilots attempted to find and remove the proper approach chart from their binders.

"...And, Tulua one...Rozo...there it is...see that comes off Tulua..."

Questions were beginning to arise as to exactly how the route should be flown. To simplify the procedure and confirm a specific point to which they could fly, the captain then decided to request a new clearance. "Can American Airlines uh, nine six five go direct to Rozo and then do the Rozo arrival, sir?" ROZO was a nondirectional radio beacon (NDB) that was located only a mile-and-a-half from the end of the runway and was a component of the approach.

The controller, whose command of the English language was limited to what was later described as "routine aeronautical communications," responded, "Affirmative, take the Rozo one and runway one niner, the wind is calm." This was the same clearance issued previously. He later reiterated, "OK, report Tulua twenty one miles and five thousand feet, American nine, uh, six five." The controller was still anticipating that flight 965 would fly the entire approach procedure, starting at the Tulua VOR. This misunderstanding that was developing between the controller and the pilot would prove to be a crucial one.

American 965 acknowledged. The first officer verified with the captain that they could descend, and started the aircraft down. Realizing that the airplane was higher at this point than it should be, he then pulled the speedbrake lever, located on the left side of the center control console. Speed brakes are large panels on the on the upper surface of each wing that are used to degrade lift and increase drag, and they remain deployed as long as the handle is in the aft position.

For the next forty seconds, the pilots discussed the approach procedures and tried to program the FMS to identify the Rozo NDB and Tulua VOR (identifier ULQ) on their navigational displays. But what was shown on their instrument panel screens was not at all what they expected to see. Their navigation displays should have been projecting a line that was more or less directly straight ahead, but instead the line curved sharply to the left. And the Tulua VOR was not shown at all.

When the controller asked for AA 965's distance from the Cali VOR, the captain replied, "OK, the distance from uh, Cali is uh, thirty eight." At about the same time, the first officer asked, "uh, where are we?"

"We goin' out to..."

"Lets go right to uh, Tulua first of all, OK?" asked the captain. "Yeah, where are we headed?" responded the first officer.

"Seventeen seven, ULQ uuuh, I don't know what's this ULQ? What the...what happened here?" "Let's come right a little bit." The captain was beginning to realize that he could not orient their position to either the Rozo NDB or the Tulua VOR.

Every moment of conversation brought the Boeing 757 closer and closer to dark, high and precipitous terrain. "...You want a left turn back around to ULQ?" asked the first officer. "Nawww...hell no, let's press on to..."

"Tulua?"

"Where we goin'?" asked the captain.

"Let's go to Cali. First of all, let's...we got [messed] up here, didn't we!" And six seconds later, "...go direct...how did we get [messed] up here?"

Sensing imminent danger and aware that Cali was off to the right, he then directed the first officer to turn the aircraft in a southerly direction, which was more generally towards the Cali airport. "Come to the right, right now, come to the right, right now!"

The captain realized that he needed verification of the route to be flown. He quickly asked the controller, "And American uh, thirty eight miles north of Cali, and you want us to go Tulua and then do the Rozo uh, to uh, the runway, right? To runway one nine?"

Because flight 965 had never called in to report passing the Tulua VOR, the controller thought that they had still not yet started the approach. But the latest report of thirty-eight miles from the Cali VOR implied that 965 had already passed Tulua and was closer to the airport than the controller expected. Confused by the latest transmission, he replied "...you can landed, runway one niner, you can use, runway one niner. What is altitude and DME from Cali?"

American answered, "OK, we're thirty seven DME at ten thousand feet."

"Roger," was the reply. "Report five thousand and uh, final to one one, runway one niner."

But the pilots were still puzzled. They were unsure of their position, of the location of various navigational fixes, why those points were not shown on the computer displays, and which procedure should be utilized.

A few seconds later, at 9:41 P.M., the controller was concerned as well. He asked "niner six five, altitude?"

"Nine six five, nine thousand feet" was the reply.

"Roger, distance now?"

There was no answer from American flight 965, nor would there ever be.

At that moment, the noise in the cockpit was deafening. The Ground Proximity Warning System's (GPWS) synthesized voice was blaring, "terrain, terrain," accompanied by the "whoop whoop" siren. The autopilot was disconnected, causing its own distinctive aural warning, and a terrain avoidance "escape" maneuver was initiated by the pilots. The throttles were slammed full forward against their mechanical stops and the nose of the airplane was raised dramatically in an effort to climb. Trading altitude for airspeed, the 757 approached an aerodynamic stall, which activated the stall warning "stick shaker," which as designed, violently shook both pilots' control columns to warn of the impending stall.

For thirteen seconds, the pilots, by then desperately aware of impending disaster, attempted to coax the aircraft over the mountain. "Up baby...up, up, up!" "More, more... pull up!" But the speedbrakes were not retracted, and the added drag sealed their fate. At 9:41.28 P.M., with the nose pitched up and the engines screaming at full power, American flight 965 impacted the side of El Deluvio.

Striking trees only 250 feet below the summit of a ridge, the aircraft started to fragment, shedding small pieces as it cut a swath through the forest. Engine cowlings and wing and tail parts were ripped from the fuselage, disappearing into the darkness below. A moment later, the body of the airplane slammed into the ground and instantly disintegrated. Momentum hurled the largest pieces over the top of the ridge, allowing them to tumble five hundred feet down the other side of the hill. Broken trees, bodies, and shards of twisted metal were thrown haphazardly across the mountainside. Smaller, lighter pieces continued to fall and fly about for a few seconds, but within a minute all movement had stopped.

When the smoke from the initial explosion cleared, only a few large pieces of wreckage were identifiable: the two engines, one of which was partially buried, the smashed cockpit, one fuselage section about thirty-five feet long, part of the tail, and a portion of the wing center section. Quiet and darkness again enveloped the mountain, but not before 159 people perished. (See Fig. 3-2.)

The rescue

Technically, this accident was classified as "nonsurvivable." That is, the impact forces and destruction of the passenger cabin were so great that no one could be expected to survive. But once in a while miracles do happen.

On the evening of December 20, Aragon International Airport was jammed with hundreds of people awaiting the arrivals of two flights, this one from Miami and a domestic Avianca aircraft. The mood was jubilant, as family and friends were returning home for the holidays.

As they waited, it was announced that American flight 965 had been delayed. When the new expected arrival time passed and there was no update, a few were concerned. After a public address announcement was made that an American Airlines agent was needed in the control tower, everyone became anxious. And when it was finally admitted that the air traffic control facility had lost all contact with the flight, pandemonium broke out in the terminal.

3-2 *Rescuers comb the wreckage of American 965. . .* Courtsey NTSB

One Colombian gentleman there to meet the arrival of his brother on flight 965 had arrived later than most. Upon hearing the news, he rushed outside to one of the many ambulances that had been called to assist in a possible rescue effort. Being a doctor, he was allowed to ride along to the small town of Buga, forty-six miles from Cali but close to where the aircraft was believed to have crashed.

The small expedition drove for several hours up into the hills, and set up a base camp on the side of San Jose Mountain. Local firemen and other would-be rescuers assisted in the search, spending the night hiking through the rugged terrain towards the summit. For protection from the leftist guerrillas that operated in the area, most of the teams were accompanied by government army patrols.

At 6:30 the next morning, a helicopter spotted the crash site. At 9:00 A.M., an experienced paramedic searching on his own came upon the devastating scene. But there, in the middle of all the carnage, were five survivors huddling in the cold. All were bloodied, dazed, and disoriented, but they were alive. Later in the day, they were airlifted out to the base camp, and there, against all odds, the doctor was reunited with his brother.

Four of the five survived, and a day later, rescue teams discovered a dog, not only alive but healthy, still in his original cargo container. They nicknamed him "Milagro" (Miracle)—the name stuck.

The investigation and findings

The investigation was conducted according to the protocols contained in Annex 13 (Aircraft Accident and Incident Investigation) of the International Civil Aviation Organization's (ICAO) International Standards and Recommended Practices. Representatives of Aeronautica Civil of the Republic of Colombia led teams of investigators from the U.S. National Transportation Safety Board (NTSB), Boeing, American Airlines, the Allied Pilots Association, and others during the on-site phase of the investigation.

One piece of very important evidence was recovered from the wreckage. The left FMC was found in operable condition and with only a small dent in the outer case. Although computer memory is normally "volatile," or deleted when power is removed from the circuit, in this case an intact backup battery kept the memory active. This resulted in the guidance buffer, built-in test files, navigation database, and operational programs all being recoverable.

Research away from the crash site

In addition to the time spent working at the site, many months were spent in research and testing at other facilities. At Boeing Commercial Airplane Group in Seattle, Washington, simulations were conducted using a B-757 fixed-base simulator to duplicate flight performance conditions and parameters. Simulations were also conducted on CDU and FMS components for system testing and aircraft interface issues. Honeywell Air Transport Systems in Phoenix, Arizona, the manufacturer of the FMCs, conducted extensive testing on the recovered electronic components.

Visits were made to Englewood, Colorado, as the Jeppesen Sanderson Co. provided both the software used for updating the on-board navigation database and the printed approach charts used by the pilots. Finally, many meetings were held with American Airlines, Fort Worth, Texas, regarding airline operations, flightcrew training, and maintenance issues.

Initial evidence indicated that flight 965 experienced no mechanical malfunctions of the aircraft, the weather was good and not a factor, the flightcrew was properly trained and qualified, and that in all respects, the flight was normal until entering Cali Approach Control airspace.

Why then, would this experienced crew allow the airplane to proceed off the planned course and continue its descent into an area of known mountainous terrain? A series of operational errors became the focus of the investigation.

The decision to accept Runway 19

The first officer had never flown into Cali, and the captain had never made this approach or landing to Runway 19. The flight had been planned for the ILS approach to Runway 01, and that procedure remained in the route page of the CDU.

The crew accepted the offer to land south in order to expedite the arrival into Cali. However, the aircraft was too high to initiate the approach, the route would need to be reprogrammed into the FMS, and neither pilot was familiar with the approach procedure. Because of their proximity to the Tulua VOR, the initial approach fix or starting point for the procedure, preparation time was very limited. Consequently, the investigation determined that the crew actions were hurried and several critical actions, such as approach procedure review, briefing, and reprogramming of the FMS, were never accomplished.

Investigators also determined that in the rush to start the descent, the captain, who was the nonflying pilot, entered commands into the CDU and "executed" them without the normal verification from the other pilot. They concluded this might have effectively removed the first officer from the decision-making and review process.

Situational awareness

A flightcrew's understanding of the position and projected flight path of the airplane in relation to navigational aids and terrain is referred to as "situational awareness." Investigators determined that one of the first mistakes made was the captain's assumption of a "direct" clearance to the Cali VOR, when in fact what was said was "...cleared to Cali," implicitly meaning via the flight plan. The controller, however, reinforced this misunderstanding by giving an affirmative acknowledgement when the captain subsequently requested verification.

When "direct" to Cali was then programmed into the CDU, all intervening waypoints, including the Tulua VOR (ULQ), dropped out of the list of active waypoints and were no longer displayed on the pilot's moving map display. Without additional action, the flightcrew could not immediately locate ULQ, resulting in their flying past the VOR without noticing it.

A few moments later the captain asked to go direct to the Rozo NDB and then to continue the arrival. Investigators believed that this indicated his lack of understanding of the approach, as Rozo is thirty miles south of the initial approach fix, and is located only a mile and a half from the end of the runway (see Fig. 3-3).

Awareness of terrain

The Aeronautica Civil determined that because the first officer had never been to Cali, he was relying heavily on the captain's experience. This may have reduced his assertiveness in questioning the flightpath of the aircraft. All thirteen of the captain's previous landings, however, had also been at night, and there is little to indicate that the crew was aware of the hazardous local terrain.

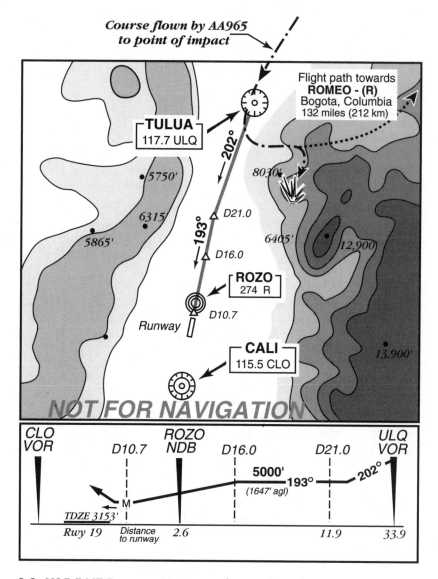

3-3 *VOR DME Runway 19 approach at Cali and American 965's flight path. . .* Source: Flight Safety Foundation

The pilots had been trained in the peculiarities of flying in South America, with additional information about certain high-altitude airports. Cali was not one of these "special" airports. No terrain information was displayed to the pilots by the FMS on the moving maps, and only high points were represented on the approach chart. A local area navigational chart did graphically depict higher elevations, but there is no indication that it was ever referenced by the crew during the descent.

The investigators also believed that the captain might have been relying on the Cali approach controller to provide terrain clearance, just as he would expect in the United States.

Automation

The use of "glass cockpit" advanced technology allows efficient flightpath management, comprehensive aircraft systems monitoring and management, and navigational flexibility and redundancy. Each pilot's moving map graphically displays aircraft position, projected flightpath, and relative positions of selected navigational aids and airports. Pilots have come to appreciate and rely upon the advanced features of the FMS.

The use of cockpit automation is not without risks, however. The Aeronautica Civil identified them as "over reliance on automation, shifting workload by increasing it during periods of already high workload and decreasing it during periods of already low workload, being 'clumsy' or difficult to use, being opaque or difficult to understand, and requiring excessive experience to gain proficiency in its use."

The report goes on to say that self-induced time constraints may have caused this crew to commit the critical error of programming and executing a course change without verifying its effect of the airplane's flightpath. The captain entered the navigational fix "R" thinking it to be the approach fix Rozo. The identifier on the approach chart is indeed "R." But to enter that particular waypoint into the FMS would require spelling the name out, as in "R-O-Z-O," since the letter "R" by itself is used by at least 12 other fixes in the airplane's onboard electronic database. Selecting the closest "R" to their present position resulted in a new course direct to "Romeo" NDB, over 130 miles northeast of Cali, near Bogota. Once this erroneous fix was programmed into the FMS, a dashed white line would have indicated the new course away from Cali. An "insufficient fuel" message should also have alerted the crew to the error. But no notice was taken, and the backlit "execute" button was pushed. Unverified prior to selection, the aircraft started a turn away from Cali (see ground track as diagrammed in Fig. 3-3).

Investigators cited the pilot's continued use of the automation in trying to locate Tulua VOR and Rozo for compounding their difficulties.

"Raw data" information, that is, tuning the VOR and NDB frequency manually and flying directly to the stations without use of the FMS, could have been done, but was not. Authorities found that the captain's inability to locate Rozo or Tulua in the FMS and the limited time available to perform other critical tasks created a very stressful situation and lack of situational awareness.

Crew Resource Management (CRM)

Investigators immediately saw similarities with another FMS-equipped aircraft accident, one involving a Thai Airways A-310 at Katmandu, Nepal. During a period of high workload, that crew experienced a system problem, which also caused a loss of situational awareness. Believing they were flying away from high terrain, they actually were proceeding directly toward it. They were confused by the navigation information displayed, and both pilots were trying to verify aircraft position using only the FMS when the airplane flew into a mountain.

American Airlines has a model CRM training program, which includes discussion of the details of the Thai Airways crash. But even with this training, crew utilization of effective CRM techniques cannot be assured during periods of stress or high workload. Specific CRM deficiencies were cited in the actions of the American Airlines' flightcrew, in the report, including:

- Using the FMS as the only source of position information was not effective, and resulted in confusion.
- Neither pilot understood how the approach should be flown, but continued inbound regardless.
- There were numerous indications that the decision to accept the approach and landing to runway 19 was a poor one, but neither crewmember questioned that choice at any time.
- Their situation was very similar to an accident scenario that had been reviewed in their recent CRM training, but went unrecognized.
- Neither pilot effectively monitored the flightpath of the aircraft for the last minute of flight.

Speedbrakes

It has never been determined if rapid retraction of the speedbrakes during the escape maneuver would have prevented impact with the terrain. It would have been close, as only an additional 250 feet was needed. However, there were few cues available to the pilots that the speedbrakes were deployed, nor was there a procedure to verify speedbrake position before executing the maneuver. Unlike some airplanes with auto retraction features, the B-757 speedbrakes will remain extended at full power settings in flight. Only if they fail to stow when commanded by the pilots will a warning sound in the cockpit.

Conclusions and probable causes

As stated in the final report issued by Aeronautica Civil only eight months later, the probable causes were determined to be:

"**1.** The flightcrew's failure to adequately plan and execute the approach to Runway 19 at SKCL [Cali, Colombia] and their inadequate use of automation.

2. Failure of the flightcrew to discontinue the approach into Cali, despite numerous cues alerting them of the inadvisability of continuing the approach.

3. The lack of situational awareness of the flightcrew regarding vertical navigation, proximity to terrain, and the relative location of critical radio aids.

4. Failure of the flightcrew to revert to basic radio navigation at the time when the FMS-assisted navigation became confusing and demanded an excessive workload in a critical phase of flight."

Contributing factors

Cited in the report as contributing to the cause of the accident were the crew's continued expediting of the approach, executing the escape maneuver with speedbrakes deployed, FMS logic that drops all intermediate waypoints from the display, and FMS naming conventions that allow identifiers in the airplane database to be different than those published in the printed navigation charts.

Also noted in the Colombian report were several other important findings. It was determined that the Cali Approach controller acted completely within International Civil Aviation Organization (ICAO) and Colombian air traffic control rules, and did not contribute in any way to the cause of the accident. Even though his English language abilities were limited, cultural and procedural differences precluded his questioning the intentions or the course flown by the flightcrew.

And while deficiencies in the Federal Aviation Administration's (FAA) surveillance of American Airlines' operations into Cali existed, specifically in the qualifications of the inspectors, those oversights played no role in the accident.

Finally, American Airlines was found to have conducted proper and adequate training in all aspects of South American flying.

Other submissions to the report

American Airlines' submission to the investigative authorities concentrated on four issues: Identifiers assigned to navigational aids, communications between the aircraft and the controller, the GPWS escape maneuver, and wreckage documentation.

Because of separate conventions for naming and identifying fixes, one for charts and one for electronic data, pilots may not be fully aware of the built-in traps in programming the FMS. The identifier used for any particular navigational fix on the printed chart, the one referenced by the pilot, may not be the same as the identifier used in the FMS. Selecting a fix in the FMS based on chart identification can lead to confusion, exactly as happened with this crew. And once they did encounter difficulties, this crew increased their dependency on the FMS and did not properly prioritize their approach preparations.

In reviewing the performance of the controller, American noted that he was concerned about the position reports he was receiving from flight 965, but that "he could not formulate his concerns into English and communicate them with the crew." Additionally, there were a number of distractions evident on the air traffic control (ATC) recording, including background music, rhythmic tapping, and non-pertinent telephone conversation at the controller's station.

The American Airlines submission further stated that the first officer performed the escape maneuver as trained, but was inhibited from attaining maximum aircraft performance because of the limited flightpath information available to him. They theorized that if the aircraft had been flown at a constant stick-shaker angle-of-attack instead of varying the pitch attitude to control stick-shaker activation, a significant performance improvement would have been available. A dedicated angle-of-attack indicator would allow the pilot to fly at the maximum coefficient of lift, thus greatly increasing performance.

Finally, American Airlines commented that the exact latitude and longitude of the initial impact point was never established. The GPS

(Global Positioning System) position obtained by the investigation team did not correlate with other known sources, and as such, the "wreckage diagram in the Structures report is abbreviated and should not be considered complete."

American Airlines believed the probable causes of the accident to be: inadequacies of the aircraft electronic database, the flightcrew's failure to note the turn away from the intended routing, and the controller's lack of understanding of the English language and his inattention during flight 965's approach. Contributing were the lack of any radar coverage, issuance of nonstandard clearances, increased flightcrew workload brought on by the change in routing, and the inappropriate overconfidence in the FMC's capabilities as demonstrated by both the pilots and the manufacturer.

The Allied Pilots Association (APA), the union that represents the pilots at American Airlines, submitted their own comments regarding the accident. Those included their contention that the engines appeared slow to accelerate to full power during the escape maneuver, and that the crew was saturated by thirty-eight radio transmissions in the last six-and-a-half minutes of the flight. APA also supported American's call for angle-of-attack indicators on the flight deck and suggested that the crew may have been under the mistaken impression that they were in constant radar contact for the entire flight.

Boeing's submission was a detailed review of the crew inputs to the FMS and the resultant aircraft flight profile.

Recommendations

Because of the investigation, the Colombia Aeronautica Civil issued seventeen recommendations to the FAA, three to ICAO, and two to American Airlines.

Those to the FAA included:

- Develop and implement standards to display terminal environment information on the FMS moving map displays, and match the portrayal of that information to what is presented on the approach charts.
- Require manufacturers to modify FMS logic to retain all intermediate waypoints when a "direct" command is programmed.

- Require airlines to train the crews as to when FMS use is an obstacle, and should be discontinued in favor of basic radio navigation techniques.

- Require that all approach charts graphically display all nearby terrain.

- Require that all pilots operating FMS airplanes have the applicable approach chart open and readily accessible.

- Encourage manufacturers to develop methods to graphically show terrain information on flight displays.

- Require Jeppesen Sanderson Co. to inform airlines of the naming conventions used, and to require the airlines to train their pilots in the logic used in creating identifiers for waypoints.

- Evaluate the present Boeing procedure for guarding the speedbrake handle during deployment, and requiring its implementation at airlines.

- Evaluate the effects of having the speedbrakes automatically stowed when high power is commanded on existing airplanes and requiring newly certified transport-category aircraft to have automatic speedbrake retraction capabilities.

- Develop mandatory Controlled Flight into Terrain (CFIT) training programs.

- Evaluate the possibility of recording pilot inputs into the FMS on the Flight Data Recorder (FDR) for more effective accident investigation.

To ICAO:

- Urge member states to strictly conform to ICAO standard radio communication phraseology.

- Consider implementing the recommendations produced by the Flight Safety Foundation's CFIT Task Force.

- Develop a single worldwide standard for all electronic database criteria.

To American Airlines:

- Standardize the evaluation criteria used by check pilots and address the analysis of flightcrew performance in training records.

Industry action

Following publication of the official accident report, the U.S. National Transportation Safety Board (NTSB) issued a number of recommendations of their own. Similar in content to those of the Aeronautica Civil, they focused on situational awareness (A-96-106), terrain depiction on aeronautical charts (A-97-102), the proper design and use of speedbrakes (A-97-90, 91, 92), FMS displays (A-96-97, 98, 99), pilot training (A-97-95), angle-of-attack indicators (A-97-94), and Enhanced Ground Proximity Warning (EGPWS) systems (A-97-101).

Because of this investigation, the following actions have been taken:

The FAA funded a research project that culminated in a report titled "Guidelines for Situational Awareness Training," published in February of 1998. The objective was to develop a program to help recognize "specific cues for situational awareness in automated cockpits," and was designed for use by both air carrier operators and FAA inspectors. In addition, appropriate guidance is to be included in a new version of AC (Advisory Circular) 121-51B, "Crew Resource Management Training."

To address the specific training issues of CFIT, the Boeing Company and the FAA developed and published the "Controlled Flight into Terrain Education and Training Aid." This program and video emphasizes accident prevention through "effective communication and decision making behavior and the importance of immediate, decisive and correct response to a ground proximity warning."

The Society of Automotive Engineers (SAE) G-10 Charting subcommittee, the Interagency Cartographic Committee- and the FAA-established FMS task force are all working to develop international standards for naming and portraying of navigational information. Once agreed upon by member nations, the International Civil Aviation Organization (ICAO) becomes the governing body and is responsible to implement necessary changes. Late in 1998 all recommendations to ICAO's Aeronautical Information Service/Aeronautical Charts Divisional Meeting were implemented. All 500 en route fixes and 3100 terminal navigation fixes will be charted as "computer" fixes in 1999.

The Jeppesen Sanderson Company had already begun publishing terrain contour information on approach charts prior to 1995. That effort continues. But there has been no recognized U.S. standard for either selection of qualifying airports or the depiction methods to be used on the charts. To that end, a Government/Industry Charting forum has been evaluating all proposed procedures, and will be recommending whether the current international standard or a new U.S. developed standard is appropriate for use.

In January of 1998, after evaluating the Boeing-recommended procedure of guarding the speedbrake handle during use, the FAA issued a Flight Standards Handbook Bulletin. Titled "Monitoring of Speedbrake Position During GPWS Recovery Maneuvers," this document directs Principal Operations Inspectors (POIs) to "ensure that their air carrier's Flight Crew Operating Manuals or flightcrew simulator programs include GPWS escape maneuver training with speedbrakes deployed." This type of training was deemed necessary because the FAA found that simply "guarding" the handle was not always operationally possible.

The FAA evaluated the feasibility of requiring both automatic stowing of speedbrakes and angle-of-attack displays on existing and newly designed transport-category aircraft. Initially, the FAA responded to both recommendations that, "if it is determined that [automatic speedbrake retraction or angle-of-attack indicators] are warranted, the FAA will take appropriate regulatory action." However, the FAA stated in June of 1999 that Terrain Awareness and Warning Systems (TAWs, including a variety of "Enhanced Ground Proximity Warning Systems," or EGPWs) hold the most promise for avoiding CFIT accidents in the future. As such, they have proposed installation of these systems in all transport category aircraft (Notice of Proposed Rulemaking, NPRM 98-11, "Terrain Awareness and Warning Systems") by four years after the effective date of the rule, or approximately 2005. Since they believe that this and other technological advances will reduce the need for ground proximity escape maneuvers in general, no modifications to existing speedbrake systems will be required. Furthermore, no change in the regulations mandating automatic retraction of speedbrakes in future aircraft design is being considered. And finally, for similar reasons, the FAA has chosen not to mandate angle-of-attack systems in transport-category aircraft.

The FAA rejected a Safety Board recommendation that bypassed waypoints remain on the FMS map display after a "direct" command by the pilot. They note that the pilot can manually input that information into the system via the NAV AID or WPT commands, although the procedure may be operationally awkward. As stated in their response to the NTSB, "The location of waypoints and navigational aids can be more simply and accurately determined by using other dedicated functions that already exist in current electronic maps' display systems."

Enhanced GPWS systems as envisioned in the FAA's proposed rule are now available. Using the airplane's onboard database and computerized flightpath profile to identify terrain and obstruction hazards, EGPWS systems may be the most valuable weapon yet available for the prevention of CFIT accidents. They are now routinely installed in new transports, and the issuance of NPRM 98-11 is just the beginning of an industry-wide effort to aggressively lower the CFIT accident rate (see Chapter 9, "Waiting for rescue," as well).

American Airlines

In conjunction with other safety initiatives, American Airlines took immediate steps to minimize CFIT risks on nonprecision approaches (see Chapter 1). These steps included:

- Additional pilot training in proper nonprecision approach procedures.
- Increasing all nonprecision approach MDAs and visibility requirements.
- Raising the ceiling of the "sterile cockpit."
- Modifying altimeter setting procedures for all flight operations.

Epilogue

One legacy of this tragic accident in Colombia has been its effect on the industry's attitudes concerning CFIT accidents. As defined in Chapter 2, "an event in which a mechanically normally functioning airplane is inadvertently flown into the ground, water, or an obstacle," CFIT accidents are the leading cause of transport-category hull-loss accidents in the world. In 1993, the Flight Safety Foundation organized a task force to study the problem, and in 1997 a comprehensive training aid was developed and published by a combined government and industry group consisting of thirty-two agencies, airlines, and associations. The Controlled Flight Into Terrain accident can only be prevented through effective flightcrew training and an industry-wide awareness of the problem. The first steps have been taken.

References and additional reading

Aeronautica Civil of the Republic of Colombia. 1996. *Aircraft Accident Report, Controlled Flight Into Terrain, American Airlines Flight 965, Boeing 757-233, N651AA, Near Cali, Colombia, December 20, 1995.* Santaf, de Bogot D.C., Colombia.

Dornheim, Michael A. 1996. Recovered FMS memory puts new spin on Cali accident. *Aviation Week & Space Technology.* 9 September, 58-61.

Federal Aviation Administration. 1998. *NTSB Recommendations to FAA and FAA Responses Report.* <http://nasdac.faa.gov/lib>

Flight Safety Foundation. 1998. Boeing 757 CFIT accident at Cali, Colombia, becomes focus of lessons learned. *Flight Safety Digest.* May, June, 1-19.

Flight Safety Foundation. 1996. Preparing for last-minute runway change, the Boeing 757 flightcrew lost situational awareness, resulting in collision with terrain. *Accident Prevention.* July, August, 1-23.

His Majesty's Government of Nepal. 1993. *Aviation Accident Report, Thai Airways International, Ltd., Airbus Industrie A310-304, HS-TID, Near Katmandu, Nepal, 31 July, 1992.* Katmandu, Nepal.

Mercer, Pamela. 1995. Inquiry into Colombia air crash points strongly to error by pilot. *The New York Times.* 29 December, A10.

National Transportation Safety Board. 1996. Public Docket. *American Airlines Flight 965, Boeing 757-233, N651AA, Near Cali, Columbia, December 20, 1995. DCA96RA020.* Washington, D.C.: NTSB.

Sider, Don. 1996. Miracle on the mountain. *People Weekly.* 8 January, 54-55.

Sutton, Oliver. 1997. New displays key to safe flying. *Interavia Business and Technology.* May, 39-41.

4

A tale of two tragedies

Birgenair 301 and Aeroperu 603

Operator: Birgenair

Aircraft Type: Boeing 757-225

Location: Caribbean Sea, twelve miles off the coast of the Dominican Republic

Date: February 6, 1996

Operator: Aeroperu

Aircraft Type: Boeing 757-200

Location: Pacific Ocean, thirty miles off the coast of Peru

Date: October 2, 1996

In order to pilot his Flyer, the first successful powered aircraft, Orville Wright lay prone on the upper surface of the lower wing. With his head into the wind, slight movements of his body and arm changed the angle of the wing and elevator, precariously maintaining "controlled" flight. None of the flight instrumentation so typical on today's modern transports was yet invented; only a stopwatch, engine tachometer, and wind speed indicator were installed on his craft. It was totally up to the perceptive skills of the pilot to recognize and respond to his airplane's height above the ground, flight-path angle, and relative airspeed. As aircraft became more complex, instrumentation—at first rudimentary and later more sophisticated—replaced most elements of "seat-of-the-pants" flying with logically developed, universally accepted operational practices and procedures.

Today's advanced airliners utilize state-of-the-art digital flight guidance systems, triply redundant autopilot and autoflight controllers, and incredibly accurate navigational equipment. Multiple air data computers are networked to collect, process, and display extraordinary amounts of information to the pilot (see Chapter 3 for additional system descriptions). Without automation, it would be very difficult—perhaps impossible—to adequately monitor and control these systems. And while the interface between the human and the automated cockpit plays a crucial role in every aspect of flight management, the pilot remains the final authority in the safe operation of his or her aircraft.

Two incredibly similar accidents in 1996 demonstrated just how important proper information interpretation and pilot response can be. Both involved Boeing 757 aircraft (see Fig. 3-1, previous chapter), foreign crews, overwater operations at night, flight guidance system anomalies shortly after takeoff, and both proved fatal to all passengers and crewmembers on board. This, then, is the tale of these two tragedies.

Birgenair, flight history and background

TC-GEN was a Turkish-owned Boeing 757-225 operated by Birgenair and leased to the small Dominican airline Alas Nacionales. Built in 1986, the airplane was primarily used for holiday charters, flying European tourists to points around the globe.

After a series of short flights that were completed on January 23, 1996, the aircraft remained on the ground in Puerto Plata, the Dominican Republic, for ten days. Maintenance had been performed and the engines run-up in the interim, but no flight tests or "return to service" checks were deemed necessary. Three days after the engine ground run, and two-and-a-half hours prior to the planned departure of flight 301, Birgenair operations was notified of a schedule change. A mechanical problem with another aircraft was discovered that would necessitate a replacement aircraft and crew for the flight to Berlin and Frankfurt. Reporting to the airport at 10:15 P.M. (local time), the substitute flightcrew verified the flight plan, checked the weather, and completed all other preflight duties. The late arrival of a flight attendant, however, caused an additional one-hour delay, and it was not until 11:40 P.M. that Birgenair flight 301 finally taxied to the runway.

With a fuel stop planned in Gander, Newfoundland, three pilots were on duty in the cockpit for the long, overnight flight. The Turkish captain, sixty-two years old, was very experienced, having accumulated almost 25,000 hours of flight time, with 2000 hours in the 757. The relief captain, fifty-one years old, had just completed training and had only 120 hours in the aircraft, although his logbook indicated 15,000 hours total flying time. One hundred and seventy-nine passengers were aboard, mostly Germans, as were ten Dominican cabin crewmembers.

The captain elected to fly the first leg of the trip, from Puerto Plata to Gander. Rolling down the runway in light rain, the first officer made the required "eighty knots" call. Just then, the captain responded "My airspeed indicator's not working!" A few moments later he asked, "Is yours working?"

"Yes sir" was the reply.

"You tell me..." commanded the captain, telling the first officer to provide him with correct airspeed information.

"Vee one, rotate," called the first officer as the plane continued to accelerate.

"Positive climb, gear up," called the captain, and the aircraft began its ascent.

Turning off the windshield wipers, the first officer asked if he should couple the flight directors with the inertial navigation systems, and was told he could do so. The captain then remarked "It [the captain's airspeed indicator] began to operate." Forty-five seconds after liftoff, the flaps were retracted, climb power set, and the after takeoff checklist was completed. The center autopilot was turned on and the crew began to settle into their normal routine. But flight 301's problems were not over, only disguised. A few seconds later, passing through 4,500 feet mean sea level (msl) altitude, the RUDDER RATIO and MACH/SPD TRIM advisory messages appeared on the Engine Indication and Crew Alerting System (EICAS) display.

"There is something wrong...there are some problems," remarked the captain. Fifteen seconds later, he again asked, "Okay, there is something crazy...do you see it?"

The first officer was also confused. "There is something crazy there! Right now mine [airspeed indicator] shows only two hundred [knots] and decreasing, sir!"

"Both of them are wrong!" shouted the captain in disbelief. "What can we do?"

Circuit breakers were checked and reset without effect. Comments made by both crewmembers implied that the alternate airspeed indicator was functioning and correct. Located on the center instrument panel, that indicator was independent of either the captain's or first officer's airspeed indication systems. A discussion between the pilots ensued, including the captain's statement that because the aircraft had been sitting on the ground for a long time some minor system anomalies such as elevator asymmetry could be expected. "We do not believe them [the EICAS advisory messages]," the captain stated. Unwilling to disagree, almost twenty seconds of silence ensued before the relief captain suggested resetting circuit breakers to "understand the reason [for what was happening]."

A few seconds later the overspeed warning clacker sounded, but the captain said, "Okay, it's no matter." He then asked that its circuit breaker be pulled to disable the noisy alarm. It was, and the warning ceased just as the first officer confirmed an indicated airspeed of 350 knots as shown on the captain's indicator. Five seconds later, climbing through 7,000 feet msl with the center autopilot commanding eighteen degrees nose up and the autothrottles at a very low power setting, the stall warning "stick shaker" activated. Its characteristic rattling sound continued until the end of the CVR recording, one minute and twenty-five seconds later.

On the flight deck confusion reigned. The autopilot remained engaged briefly, although the autothrottles automatically disconnected and the VNAV (vertical navigation) function of the auto flight system was no longer coupled. Exclamations of surprise and dismay were exchanged. Aircraft pitch attitude oscillated between five and twenty one degrees (aircraft nose up, ANU), full power was applied and then immediately removed. The airplane fell into a steep descent.

"We are not climbing? What am I to do?" the captain cried out.

"You should level off...I am selecting altitude hold, sir!"

"Select...select!"

Without power and under manual flight, however, autopilot command inputs had no effect on the 757, and it continued to dive. "Thrust, thrust, thrust, thrust!" commanded the captain. Realizing that power levels were reduced, he again pleaded "Thrust...don't pull back, don't pull back, don't pull back!"

Once again, full power was restored to both engines. Inexplicably, however, left engine thrust was immediately reduced to a lower level. Descending through 3,500 feet a few seconds later in a fully developed aerodynamic stall, the nose of the 757 pitched down almost eighty degrees. Sensing the extreme rate of descent and the impending impact with the water, the Ground Proximity Warning System (GPWS) began to blare its unrelenting "Whoop, whoop, pull up, pull up!" warning.

"Sir, pull up!" yelled the third pilot in the cockpit. "Oh! What's happening?"

Eight seconds later, fourteen nautical miles northeast of Puerto Plata, Birgenair 301 slammed into the Atlantic Ocean. The aircraft disintegrated, instantly killing all one hundred and eighty-nine people aboard. A few light pieces floated for some time, but eventually most of the wreckage came to rest on the ocean floor, a mile-and-a-half below the surface.

The investigation and findings

Under the provisions of Annex 13 to the Convention on International Civil Aviation, the Junta Investigadora de Accidentes Areos (JIAA) of the Director General of Civil Aeronautics (DGAC) of the Dominican Republic conducted an investigation into the causes of the accident. Technical assistance was requested from the United States, and a team of specialists from the National Transportation Safety Board (NTSB) traveled to the crash site. Also included were representatives from the Federal Aviation Administration (FAA), the Boeing Company, as the manufacturer of the airplane, and Rolls Royce, as the manufacturer of the engines. Assisting the JIAA were investigators from Turkey, Germany, and the airline, Birgenair.

A Dominican tugboat and a commercial oil tanker were on the scene almost immediately and were assisted by aircraft from the U.S. Coast Guard. After finding no survivors, initial efforts were directed toward pinpointing the debris field and recovering the cockpit voice recorder

(CVR) and flight data recorder (FDR). Underwater locator signals from both units were detected on the 15th of February inside the territorial waters of the Dominican Republic, but at the incredible depth of 7,200 feet. It was soon apparent that retrieving the recorders would be a challenging and expensive undertaking, but in the interests of safety, all parties agreed to an unusual plan to share the costs, estimated to be $1.4 million (U.S.).

The United States Navy's Supervisor of Salvage (SUPSALV) and technicians from Oceaneering Technologies were called in to devise a recovery plan. The U.S. Motor Vessel Seaward Explorer mapped the ocean floor with a side-scanning sonar system and determined that the debris field was compact, only about 900 feet long and 1,500 feet wide. A third underwater beacon was placed between those of the FDR and CVR to aid in quickly locating the recorders, and the commercial motor vessel Marion C 2, equipped with deep-sea vehicles, arrived on station February 27th.

The next day, CURV III, a 13,000-lb. "cable controlled underwater recovery vehicle" was used to explore the wreckage. Thorough documentation was possible by using the unmanned submarine's closed-circuit television system and still cameras. Maneuvering easily through the depths, its titanium robotic arms deftly probed the crushed fuselage of the airplane, and after only two hours below the surface, both flight recorders were retrieved (see Fig. 4-1). The units were carefully dried out, treated with an alcohol solution to prevent oxidation, and transported to the laboratories of the NTSB in Washington, D.C. for read-out.

Investigators determined that both recorders functioned normally during their final flight, but it quickly became evident that the calibrated airspeed (CAS) values recorded on the FDR did not correlate with all the other data. Actual ground speeds of the aircraft shortly after takeoff until the final loss of control were much less than the airspeeds indicated on the captain's indicator, although close to the values read by the first officer. The two EICAS advisory messages received by the crew also pointed to an airspeed indication problem. Showing that the recorded airspeeds were incorrect was easy for the investigators; determining what caused the erroneous indications would be more difficult.

The pitot/static system

All aircraft utilize a pitot/static system to determine and display airspeed and altitude. An externally mounted "pitot" tube has a

4-1 *The CURV is raised from the ocean after recovering the CVR*

... Courtesy FAA

very small opening in the end facing into the airstream that measures "ram" air pressure. A static port, a very small opening on the outside of the fuselage, measures the ambient or nonram air pressure. In advanced airplanes, an air data computer (ADC) compares the two values and generates an electrical signal that is sent to the primary airspeed indicators in the cockpit (see Fig. 4-2). That signal is then converted into rotary movement of the pointer, which the pilot then interprets as airspeed. To assure redundancy, most large aircraft have several pitot/static systems, each independent of the other.

The left ADC in the Boeing 757 drives the captain's airspeed indicator, the center autopilot, and the airspeed recorded on the flight data recorder. It was becoming obvious to investigators that the answer to the loss of flight 301 would be found in the left pitot/static system. In fact, the sequence of abnormal indications was absolutely characteristic of a well-known problem: total blockage of the pitot tube by some foreign object.

If a pitot tube was completely blocked, no ram air pressure would be present during the initial takeoff roll, and no airspeed at all would be indicated. As the aircraft climbed, the blocked pressure in the pitot line would be interpreted as increasing (compared to ambient pressure, which was decreasing with altitude), thus generating an increasing airspeed indication. In essence, the airspeed indicator was now functioning like an altimeter, increasing its displayed reading with an increase in altitude. The FDR would record the higher values, and the center autopilot would fly the airplane also based on the inaccurate airspeeds. Investigators believed that this, in fact, is exactly what may have happened to Birgenair 301.

Flight simulation
Using a Boeing 757 flight simulator, investigators were able to create a scenario comparable to the accident flight. With the left upper pitot tube (captain's) "blocked" by simulated ice, indications in the cockpit were identical to flight 301. As the aircraft (simulator) climbed, indicated airspeed began to increase. Acting appropriately based on signals sensed by the air data computer, the center autopilot increased pitch, and the autothrottles reduced power to counter the ever-increasing airspeed. Before long, the captain's erroneous airspeed indication generated an overspeed warning. Shortly thereafter the stick-shaker also activated, and without proper flight control inputs, the aircraft (simulator) stalled.

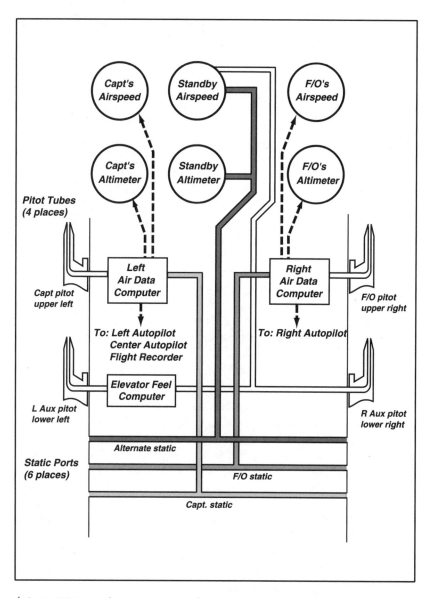

4-2 *B-757 pitot/static system schematic and selected electronic components. Dashed lines represent electrical connections...* Source: Boeing

Maintenance practices

Normal maintenance practices require that pitot tube, static port, and engine covers be installed on an aircraft whenever it is out of service for an extended period. As stated in the DGAC's final report, "During this period [not flying for the previous twenty days]

an engine inspection was performed that required an engine ground test before the next takeoff. Investigators believe that the engine and pitot covers were not installed before or after the engine ground test." Furthermore, no preflight functional verification of the pitot/static system was done as recommended by the manufacturer. "If this inspection had been completed as a part of the return to service, it may have discovered the blocked pitot tube system and it would have been corrected before the flight."

The exact reason for the blockage in this case could not be determined. As stated in the report, "...the authorities of the investigation concluded that the probable source of obstruction in the pitot system was mud and/or debris from a small insect that was introduced into the pitot tube during the time the aircraft was on the ground in Puerto Plata."

Pilot actions

The lack of indicated airspeed on the captain's instrument during the takeoff was the first indication of a problem. Investigators found that "...the captain underestimated the importance of the discrepancy between airspeed indicators experienced during takeoff, as a result of the apparently correct indication during the initial phase of the ascent." There was no specific airspeed discrepancy warning on the 757, but a ten-knot difference between indicators for five seconds will cause the MACH/SPD TRIM and RUDDER RATIO messages to be displayed. According to the report, when those EICAS alerts were triggered passing through 4,300 feet, "...the crew did not attempt to clarify the alerts or to take corrective action."

The first officer realized that his indicated airspeed was decreasing rapidly, but confusion in the cockpit interfered with the analysis of the problem and selection of the proper course of action. When the captain stated that both indicators were incorrect, he also mentioned that the "Alternate [airspeed indicator] is correct." Both pilots affirmed this critically important fact, but as the report states, "...they did not seem to understand the importance of comparing the three indicators." Neither did they "suggest...switching the instrument selector for 'alternate source' to 'alternate' to derive airspeed information from the ADC of the first officer..." That action would have caused the captain's airspeed indicator to display the same speeds as were shown on the first officer's.

The JIAA documented five sources of velocity information that would have been available to the crew. The two primary (captain's and first officer's) indicators and the alternate indicator provide airspeed information. The airplane's ground speed (derived directly from the navigational Inertial Reference Units, totally independent of the pitot/static systems) was also continually displayed on both pilot's EFIS screens; at no time during the flight was any reference made to that critical information. The report found that "The failure of the flightcrew to realize the right course of action and to understand the reduction of displayed ground speed information in the EFIS screens indicated a lack of knowledge of the aircraft systems and a lack of Crew Resource Management (CRM) in the cockpit."

Duplicating the accident flight profile in the simulator, it was determined that recovery from the stall was possible utilizing maximum power settings and appropriate and timely flight control inputs. The Boeing Company confirmed that during a developmental test flight of the aircraft, their test pilots entered a very similar stall regime but were able to regain control using normal stall recovery techniques.

Conclusions and probable cause

The JIAA determined that the weather, communications procedures, and navigational aids played no role in the accident. Postmortem examinations of the seventy-two bodies recovered and underwater wreckage inspection revealed no evidence of preimpact fire or smoke. But other issues were considered. The crew may not have been fully rested due to their unexpected call-out, the flightcrew's training did not include any CRM, and pilot training conducted at academies not affiliated with Birgenair may not have provided an "integrated" approach to in-flight problems or addressed airline-specific issues. Further, the report noted that the operations manual of the Boeing 757/767 did not contain appropriate, detailed information regarding flight with airspeed discrepancies or an "untrustworthy" airspeed indicator, nor was there a warning generated by the 757/767 EICAS system if an erroneous airspeed signal was detected.

As stated in the final report, "The probable cause of the accident was the failure on the part of the flightcrew to recognize the activation of the stick-shaker as an imminent warning of an entrance to aerodynamic stall and their failure to execute proper procedures for recovery of the control loss. Before activation of the stick-shaker, confusion of the flightcrew occurred due to the erroneous indication of an increase in airspeed and a subsequent overspeed warning."

The obstructed pitot tube, while not the probable cause, was deemed "a contributing factor."

Recommendations

The JIAA and the NTSB, as "full partners" in the investigation, issued the following recommendations to the International Civil Aviation Organization and to the Federal Aviation Administration:

- Issue a directive requiring a revision to the Boeing 757/767 flight manual advising pilots that simultaneous activation of the MACH/SPD TRIM and RUDDER RATIO alerts may be an indication of airspeed indicator discrepancies. Additional information should also be placed in the emergencies procedures section regarding "Identification and elimination of an erroneous airspeed indication." A Flight Standards Information Bulletin (FSIB) should be issued to all operations inspectors assuring that these procedures are contained in each airline's manuals. (A-96-15 and 17)

- Require the Boeing Company to modify the 757/767 alert system to include an erroneous airspeed advisory "caution" alert. (A-96-16)

- Issue an aeronautical information bulletin notifying inspectors of the circumstances surrounding the accident, emphasizing the importance of recognizing a malfunctioning airspeed indicator during takeoff. Additionally, ensure that inspectors verify that all 757/767 operator's manuals include detailed emergency procedures addressing this particular malfunction. (A-96-18 and 19)

- Assure that all Boeing 757/767 simulator training includes a "blocked pitot tube" scenario. (A-96-20)

The JIAA also recommended that ICAO (International Civil Aviation Organization) ensure that:

- Each flightcrew academy have a training program specific to the contracting airline's requirements, and that all commercial airlines and training establishments include a cockpit resource management (CRM) program.

Industry action

After careful review, the FAA decided against a mandatory revision to the Boeing 757/767 Airplane Flight Manual to alert pilots to specific airspeed discrepancies. They pointed out that advisory materials, including FAA takeoff safety training aids and B-757 Operations Manual normal procedures emphasize the importance of early recognition of erroneous airspeed information. It was also noted that the simultaneous illumination of both the MACH/SPD TRIM and RUDDER RATIO advisories can signal four different situations; the proposed revision to the manual might actually mislead a flightcrew as to the true cause of the warning.

However, as an alternate action, the FAA did issue a FSIB to Air Transport, 96-15, "Boeing 757/767 Aircraft Airspeed Indicator Malfunction Procedures and Training," on September 26, 1996. That bulletin directed principal operations inspectors to ensure that their carrier's manuals "include appropriate emergency procedures for identification and corrective action when erroneous airspeed indications are observed or suspected."

The FAA chose not to require that Boeing modify the EICAS to provide a "caution" alert when an erroneous airspeed indication is detected. Boeing, however, will consider such an airspeed "advisory" alert with upgrades to the EICAS to be certified in the future. In November of 1996, Boeing revised the B-757/767 Operations Manual to include "detailed information for recognizing an unreliable airspeed /Mach indication and guidelines for responding to the condition."

On February 27, 1998, the FAA issued change 2 to Advisory Circular (AC) 120-51B, "Crew Resource Management," which incorporated a training scenario requiring appropriate responses to the effects of a blocked pitot tube. This training was done in conjunction with Line-Oriented Flight Training (LOFT) for flightcrews, and required that "emphasis be placed on inquiry/advocacy/assertion, situational awareness and crew coordination" anytime flight instruments act abnormally.

As appropriate and timely as these actions were, though, it was not enough.

Aeroperu, flight history and background

Sadly, history has a strange way of repeating itself. Only eight months after the fatal crash of Birgenair 301 and 2,400 miles to the southwest, routine maintenance was being completed on another Boeing 757 in preparation for its scheduled international flight later that night. Like their Turkish counterparts, this Peruvian flightcrew would be confronted with erroneous and contradictory airspeed and altitude information just moments after takeoff. The modern, automated instrumentation and displays had never failed them before— would they be able to analyze and comprehend their dilemma, coordinate a plan of action, and revert to basic flying skills quickly enough to avert disaster?

N52AW was new by aviation standards. Built in 1992, it was one of only two B-757-200s in Aeroperu's fleet, leased from Ansett World-wide Aviation Services and delivered to the airline in 1994. Having logged only 22,000 hours since its manufacture, all maintenance was performed under contract by Aeromexico.

Wednesday evening, October 2, 1996, was pleasant in Lima, with scattered clouds, mild temperatures in the 60-degree F range and light winds. Mechanics had been hard at work all evening on the airplane, replacing engine inlet guide vanes that had been damaged by bird ingestion. The right engine-driven hydraulic pump was also changed, in accordance with the manufacturer's recommended life limits. The aircraft was washed and the forward part of the lower fuselage, which was bright aluminum, was polished. That activity had been interrupted, however, with a mandatory test run of the engine. The maintenance team was almost finished when the fifty-eight year old captain of flight 603 performed his "walk-around" inspection before departure, and the final paperwork was quickly completed.

Both flight deck crewmembers were very experienced, accumulating 30,000 hours of flying time in their combined careers. After assisting the captain with all flight planning duties for their trip to Santiago, Chile, it was decided that the forty-two year old first officer would be the flying pilot that evening. With sixty-one passengers and seven flight attendants, Aeroperu 603 took off at forty-one minutes after midnight.

Just seven seconds after raising the landing gear, the first sign of trouble appeared. "The altimeters are stuck!" exclaimed the first officer. "All of them!"

Just then a "Wind shear, wind shear, wind shear!" warning blared from the cockpit speakers. Surprised, the captain asked, "What's happening, we're not climbing?"

"The speed...Vee two plus ten...hold it, no...keep the speed!" Not only were the altimeters not responding normally to the airplane's climb, but the two airspeed indicators were erratic and in disagreement.

A warning appeared on the EICAS. "Rudder ratio!" the first officer called out, one minute after liftoff. "How strange," was the captain's reply, "turn to the right." Having taken off to the south, a turn to the right would take the aircraft out over open sea and away from mountainous terrain to the east.

"Go up, go up, go up!" implored the captain.

"I am!" yelled the copilot, "but the speed..."

The captain verbalized the first indication of the airspeed problem. "But it's stuck...mach trim, rudder ratio...now you are..."

Thirty seconds later, beginning to realize the seriousness of the situation, the pilots decided not to engage the autopilot. "The speed, let's go to basic instruments, everything has gone!" With caution and alert alarms sounding in the background, Aeroperu 603 declared an emergency at 12:42:32 A.M., two-and-a-half minutes after takeoff.

The altimeters then indicated 1,700 feet mean sea level (msl), which the controller confirmed as what was displayed on his radar. Throughout the remainder of the flight, the crew requested and received altitude information from ATC, using it to confirm what was shown on their altimeter. The altitude that was displayed on the controller's screen was not independently generated, but was electronically sent to ATC by the airplane's transponder, which read it directly from an aircraft altimeter. The altitude shown on the controller's radar screen, therefore, would always be the same as that shown on the aircraft altimeter, whether accurate or not.

Shortly thereafter, the captain took over all flying duties, but couldn't determine if the autopilot was on. "Autopilots have been connected," he stated.

"No, no, they are disconnected!" argued the first officer. "...only the flight director is on."

Vectors were issued by ATC to keep the flight out over the ocean while the crew tried to understand the nature of the problem. "Alternate" static source selections were made, with no changes noted to the instruments. The first officer made many attempts to find a remedy in the aircraft flight manual, reading, "...avoid large or abrupt rudder inputs, if normal left hydraulic system pressure available...yes, crosswind limit, do not attempt an autoland, it says..."

"Yeah, we can't even fly!" retorted the captain.

In the confusion, procedures normally completed immediately after takeoff were forgotten. Eight minutes after departure the flaps were finally raised and climb power set. For several minutes, the crew discussed whether or not to use the autopilot and autothrottles, and a minute later the air traffic controller called again. "Aeroperu six zero three, you are forty miles from Lima and according to my screen are level at one two zero, approximate speed over the ground is three hundred ten knots." Significantly, since the ground speed information offered by the controller did not come from the airplane's transponder but directly from computations of distance and time calculated by the ATC computer, it was a completely independent source of information.

Both crewmembers acknowledged the controller's statements, reinforcing their misconceptions of the airplane's true altitude. They also confirmed "...maintaining speed, we have two thirty..." not noticing the discrepancy between their indicated airspeed and the ground speed reported by ATC.

Again, the crew reread the flight manual, and again they found the recommended procedures of no help. While being vectored to return to the airport at Lima, the captain's indicated airspeed increased to 320 knots and the overspeed warning sounded, startling the crew. Initially believing the alarm to be legitimate, engine thrust was reduced and the speedbrakes were extended in the mistaken attempt to slow the aircraft. Lima Control again stated 603's position, "You are crossing the two six zero [degree radial] of Lima, thirty one miles west. Flight level is one hundred plus seven [10,700 feet] and approximate speed is two hundred eighty over the ground."

"Yeah, but we have an indication of three hundred fifty knots here!" responded the first officer.

The captain, his frustration complete, yelled, "I have speedbrakes, everything has gone! All the instruments have gone, all of them!"

Seventeen seconds after the overspeed clacker was heard, the stall warning stick-shaker activated. Shouting over the din of the simultaneous alarms, the first officer pleaded to the controller, "...is there any airplane that can take off to rescue us? Any plane that can guide us? An Aeroperu that may be around?...somebody?"

Just as they were informed that a 707 would be departing shortly to assist them, the cockpit was filled with yet another aural warning, this one even more ominous. "Too low, terrain! Too low terrain!" admonished the mechanical voice.

"What happened?" demanded the captain in disbelief.

"We have a terrain alarm! We have a terrain alarm!" screamed the first officer into the radio.

"Too low terrain!" continued the voice, somewhat drowning out the overspeed clacker and stick-shaker.

Airspeed discrepancies again triggered another familiar warning. "Wind shear! Wind shear! Wind shear!" added to the general clamor.

"Too low terrain! Too low terrain!"

"Wind shear! Wind shear! Wind shear!"

"All of the computers are crazy here!" radioed the first officer. "We have a terrain alarm and we are supposed to be at ten thousand feet?"

"According to the [radar] monitor you are at ten five [10,500 feet]" responded the controller.

Realizing that the airplane had been in maintenance that day, the captain commented to no one in particular, "What the hell have those [mechanics] done?"

Flight 603 turned to the west again, unsure of their position in relation to the mountains and seeking the safety of the open sea.

"We are over water, aren't we?" the crew asked "...we have like three hundred seventy knots, are we descending now?"

"Affirmative, over the water, you are forty two miles west," answered the controller. "It shows the same speed...your speed is two hundred, approximately."

"Speed is 200? We will stall now!"

"Sink rate! Sink rate! Sink rate! Sink rate!" cautioned the mechanical voice in the cockpit.

Power was added and the aircraft climbed slightly, but their altimeters continued to indicate 10,000 feet. A stall was avoided but the warnings persisted. Reading the emergency procedures guide one more time was of no help.

"OK, let's go back," the captain announced. Assuming that the situation was as stable as it was ever going to be, he wanted to attempt an approach and landing in Lima one more time.

"Do you have our speed?" the crew asked Lima Control.

"Yes, it shows a speed of three hundred knots," answered the controller.

"[The] horizons are OK...horizons are right, it's the only thing that is right!" remarked the first officer.

The flight was again vectored to intercept the ILS at Lima, and for several minutes, the crew discussed the few alternatives available to them. Checklists are reread, but no other actions are taken. Once more, they mistakenly attempted to verify their altitude with ATC. "Can you tell me the altitude, please, because we have the climb that doesn't..."

"Too low terrain! Too low terrain!" interrupted again.

"Yes," radioed the controller, "you are still at nine seven hundred [9,700 feet] according to my presentation, sir."

"Nine seven hundred? But it indicates 'too low terrain'! Are you sure you have us on the radar at fifty miles?" The first officer, believing the altitude information supplied by the controller to be correct, was now questioning their position. His assumption was that the GPWS warning was due to their proximity to mountainous terrain, not because of height above the water.

But their time was running out. Twenty-eight seconds later the 757's left wing and engine sliced through the surface of the Pacific Ocean. "We are impacting the water!" screamed the first officer into the radio. "Pull it up!"

Struggling to maintain control, the pilots flew the crippled airplane for another seventeen seconds. "I have it! I have it!" yelled the captain as he wrestled the control column. But lethal damage had been inflicted, and the aircraft continued its slow roll to the left. Thirty-one minutes after takeoff, Aeroperu 603 smashed into the sea in a very steep left bank, killing all seventy persons on board. At impact, the captain's instruments showed an altitude of 9,500 feet and an airspeed of 450 knots.

The investigation and findings

The Peruvian Accident Investigation Commission (AIC) conducted an investigation into the causes of the accident in accordance with Annex 13, assisted by investigators from the NTSB. Peruvian Navy vessels located the primary debris field and a U.S. Navy explosive ordinance team on maneuvers in the area detected the underwater locator "pingers" on the CVR and FDR the next day. Once again, Oceaneering Technologies was brought in to conduct an intensive undersea search. Using a "Hydra Magnum" remotely operated undersea vehicle deployed from the Peruvian vessel "MV Hippo," initial emphasis was placed on the recovery of human remains. Subsequently, the 3,500-lb. unmanned submarine located both recorders and retrieved them from the ocean floor, 700 feet below the surface. After sending the devices to Washington, D.C. for readout, a thorough photographic documentation of the wreckage was accomplished. One picture taken by the Magnum was particularly telling; a piece of masking tape covered all three static ports on the left side of the fuselage (see Fig. 4-3).

Blocked static ports (see Fig. 4-1 for system schematic) can cause completely erroneous airspeed and altitude information to be displayed on the pilots' instruments. If the obstruction is only partial, delays or "lags" in the display will occur, resulting in a lower than actual altitude being displayed while climbing, and a higher altitude displayed when descending. Peruvian AIC and NTSB investigators believed that the erroneous indications recorded on the CVR were consistent with a partial blockage of the static ports. It was never determined conclusively if the right side static ports were obstructed as well.

Investigators discovered that during the washing and polishing of the aircraft, masking tape was applied to the static ports to prevent the introduction of moisture and contaminants into the system. The approved maintenance procedures call for the use of "moisture-resistant" paper, however, not tape. To assure appropriate quality control, the first inspection should have been conducted by one of the

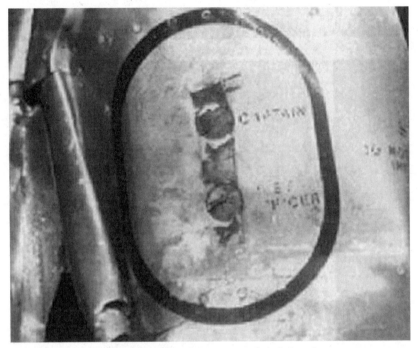

4-3 *Underwater photograph of Aeroperu 603 left side static ports. Note dark masking tape remnant covering ports.*

technicians immediately after completion of any maintenance work. A shift supervisor should then have reviewed the work, a line maintenance chief would sign it off, and in this case, the pilot would have the last opportunity to observe the tape on the static ports as he did his preflight inspection. Investigators were unable to determine why not one person in this safety "chain" noticed that the tape had not been removed after the polishing job was terminated.

Flight crew performance during the emergency was also scrutinized. Investigators realized that multiple alarms and conflicting altitude and airspeed indications caused the pilots tremendous psychological stress. As a result of their anxiety and confusion, investigators believed that the crew failed to prioritize the various problems and possible solutions, did not respond appropriately to the GPWS indications, neglected to complete any normal or emergency checklists, and never utilized proper Crew Resource Management (CRM).

The CVR confirmed that the EFIS ground speeds were never referenced, nor were the properly functioning radio altimeters. Although the crew was saturated with erroneous information and distracting aural warnings throughout the last twelve minutes of flight, no clear, decisive actions were ever taken to either identify the real problem or to remedy the situation. The AIC pointed out that while the failure to remove the tape from the static ports was an "active" failure, subsequent flightcrew errors in "situational awareness, navigation, calls, procedures and tactical decisions" were all typical of a Controlled Flight Into Terrain (CFIT) accident.

Conclusions and probable cause

The AIC determined that the weather was not a factor, that all ATC involvement was appropriate, and that the airplane loading, performance, and certification were in accordance with accepted standards. The pilots were properly certified, and with the exception of the static system, there were no aircraft or engine anomalies that contributed to the accident.

The Accident Investigation Commission concluded that the principal probable cause was "Employee error, including the crew." It was found that "...[an] employee did not remove the adhesive tape from the static ports...[nor was it] detected by any number of people doing inspections or [during] the preflight inspection performed by [the captain]."

Contributing causes were "Error of the pilot...[for] not complying with procedures for GPWS alarms..." and error of the first officer, for "...not insisting, being assertive or putting more emphasis on convincing the pilot of the [importance of the] terrain proximity alarms."

Recommendations

The Commission issued a number of safety recommendations, mostly to the operator, the maintenance crew, the Director General for Air Transport (Peru, DGTA), and to the FAA. They included:

- Provide better training to flightcrews in dealing with airspeed anomalies, to develop effective procedures for reacting to GPWS warnings, emphasizing the proper use of and confidence in the radio altimeter and more complete aircraft systems knowledge.
- Design and mandate the use of a more effective and visible static port protector.
- Implement a more effective system of quality control, maintenance technician selection, training, and retention.

- Assure strict compliance with all of the manufacturers' recommended maintenance practices.

- Create an effective safety information system and training program within the DGTA, and to strictly enforce existing regulations.

- Redesign the Boeing 757 to eliminate contradictory warnings, and provide a "caution" alert when speed or altitude information is not reliable.

Industry action

Even before the Commission of Accident Investigation issued the final accident report, the NTSB had issued its own urgent recommendation. As stated by Chairman James Hall, "The use of adhesive tape coverings alone does not attract adequate attention to ensure that the coverings are removed before flight." The letter to the FAA continued, "The Safety Board believes that the FAA should take immediate action to review and amend, as needed, all airplane maintenance manuals to require operators to use only standardized, highly conspicuous [static port] covers with warning flags attached." (A-96-141)

The FAA acted quickly on the issue, publishing Flight Standards Information Bulletin for Airworthiness 97-08A (FSIB), "Procedures to Ensure Covers are Removed From Static Ports Following Cleaning and Maintenance" on April 2, 1997. This bulletin directed Principal Operations Inspectors (POIs) to review their carrier's maintenance manuals for adequate safeguards and procedures for installation of appropriate static port covers during maintenance and cleaning. The bulletin also directed POIs to "recommend that their operators adopt the Board's recommendation to use conspicuous covers with warning flags attached in any situation in which static ports may need covering."

The NTSB was not satisfied that the actions taken by the FAA were fully responsive to the intent of the original recommendation, and urged that "POIs be directed to require, rather than recommend, use of conspicuous static port covers." In September of 1997, the FAA superseded their original bulletin with a Flight Standards Handbook Bulletin for Airworthiness (HBAW) 97-15, directing POIs to require carriers to use "conspicuous [static port] covers with warning flags," as part of any FAA-approved continuous maintenance program.

Personal liability laws differ from country to country; late in 1997 a Peruvian State prosecutor charged five Aeroperu workers, including the head of the quality control department and the chief of the paint shop, with manslaughter. Charges could still be brought against others, including air traffic controllers, if the ongoing court case determines that they were partially responsible or negligent in any way.

Epilogue

Over eight hundred Boeing 757s are in use in countries around the world, maintaining a truly enviable safety record. As a highly automated aircraft, however, the interaction between man and machine is critical—as these two accidents have shown us, any breakdown in that interface can be disastrous.

The automation that has been designed into the aircraft is incredibly reliable. So much so that pilots (or for that matter, operators of any dependable automated system) tend to become complacent, relying on the system to perform as it always has. That can lead to a reduced awareness of the functional and operational status of the system and reduced effectiveness of the operator. Paradoxically, as technology improves, once-simple tasks can become much more complex. Previously straightforward problems become harder to analyze and their solutions can become less evident.

The pilot will always be the final link in the safety chain. But to be successful in the future, the basic flying skills inherited from Orville Wright must not be replaced, only augmented, with effective technology management.

References and additional reading

Aarons, Richard N. 1997. For want of a red ribbon. *Business & Commercial Aviation*. June, 110.

Federal Aviation Administration. 1999. *NTSB Recommendations to FAA and FAA Responses Report*. <http://nasdac.faa.gov>

McKenna, James T. 1996. Peru 757 crash probe faces technical, political hurdles. *Aviation Week & Space Technology*. 7 October, 21.

——. 1996. Blocked static ports eyed in Aeroperu 757 crash. *Aviation Week & Space Technology*. 11 November, 76.

National Transportation Safety Board. 1996. Public docket. *Birgenair, Flight 301. Boeing 757-225, TC-GEN, near Puerto Plata, Dominican Republic, February 6, 1996. DCA96RA030*. Washington, D.C.: NTSB

Neumann, Peter G. 1996. Forum on risks to the public in computers and related systems. *The Risks Digest, ACM Committee on Computers and Public Policy, May 7, 1996*. 18(10). <http://catless.ncl.ac.uk/Risks/18.10>

Phillips, Edward H. 1996. NTSB seeks clues to Dominican crash. *Aviation Week & Space Technology*. 4 March, 33.

——. 1996. Airspeed discrepancy focus of Birgenair probe. *Aviation Week & Space Technology*. 11 March, 34.

——. 1996. Pitot system errors blamed in 757 crash. *Aviation Week & Space Technology*. 25 March, 30.

——. 1996. Birgenair crash raises airspeed alert issues. *Aviation Week & Space Technology*. 10 June, 30.

——. 1996. NTSB urges change in static port covers. *Aviation Week & Space Technology*. 2 December, 33.

Robles, Ricardo, Presidente, Commission of Accident Investigations (CAI), Director General of Air Transport (DGAT) of Peru. 1996. Final report, *Boeing 757-200 Accident, Aeroperu, October 2, 1996*. Lima, Peru.

Souffront, Emmanuel T., Presidente, Junta Investigadora de Accidentes Aéreos (JIAA) of the Director General of Civil Aviation (DGAC) of the Dominican Republic. 1996. Aircraft accident information. *Dominican Republic Press Release—Factual Information, March 1 and March 18, 1996*.

Souffront, Emmanuel T. Presidente Junta Investigadora de Accidentes Aéreos (JIAA) of the Director General of Civil Aviation (DGAC) of the Dominican Republic. 1996. Final aircraft accident report. *Birgenair Flight ALW-301, February 6, 1996*. Santo Domingo, Dominican Republic.

5

Carnage in the Everglades

The flight of ValuJet 592

Operator: ValuJet Airlines
Aircraft Type: Douglas DC-9-32
Location: Everglades, near Miami, Florida
Date: May 11, 1996

With scattered clouds, a light breeze out of the east, and temperatures in the mid-80s, this Saturday had all of the makings of a beautiful spring day in south Florida. As usual for a weekend, there was a bustle of activity at the Miami International Airport as scores of cruise ship passengers made connections home, while other travelers hurried to college graduations or Mother's Day celebrations.

The ValuJet ticket counter was busy as well. One hundred and five passengers had boarded flight 592 to Atlanta but were forty minutes late departing the gate due to mechanical complications with the aircraft earlier that morning. As their flight taxied for takeoff, perhaps some passengers worried that their late departure might mean inconveniencing awaiting family members, or even necessitate an annoying change of plans. But as fate would have it, they would be much more than just late—for in a cargo compartment just below the cabin floor was a lethal, smoldering brew, steeped in incredibly high temperatures and fueled by extraordinary carelessness.

Flight history and background

Flight 592 was scheduled to depart Miami International Airport (MIA) at 1:00 P.M. However, the aircraft that was to make up the flight, N904VJ, was late arriving from its previous station. Earlier that morning the flightcrew encountered a minor maintenance problem, and by the time the aircraft arrived in Miami, it was already ten minutes past flight 592's scheduled departure time.

As ground preparations were underway to quickly prepare the aircraft for its departure to Hartsfield Atlanta International Airport (ATL), the captain reported another mechanical problem—the public address (PA) system had become inoperative. ValuJet employed no company mechanics in MIA, so a contract mechanic was called in to investigate the problem. He determined that an amplifier was loose in its mount, reseated it, and then confirmed that the PA was operable.

The DC-9-32 (see Fig. 5-1) was loaded with 4,109 pounds of cargo, which included baggage, mail, and company-owned material (CO-MAT). The shipping ticket for the COMAT indicated that the company materials consisted of two main aircraft tires and wheels, a nose tire and wheel, and five boxes that were described as "Oxy Cannisters—'Empty'" (sic). As the first officer made his preflight exterior walk-around inspection of the aircraft, the ValuJet lead ramp agent showed him the shipping ticket and asked for approval to load the shipment in the forward cargo compartment. Although the two did not discuss the notation concerning the oxygen canisters, the cargo was loaded, and shortly thereafter the forward cargo compartment door was closed.

Flight 592 pushed back from the gate at 1:40 P.M. and approximately four minutes later began taxiing for departure. Using their official ATC call sign of "Critter 592," Miami Tower cleared them for takeoff on Runway 9L, about three minutes after two in the afternoon.

At the controls of flight 592 was Captain Candalyn "Candi" Kubeck, a skilled pilot who had celebrated her 35th birthday one day earlier. Beginning flying in high school, Captain Kubeck pursued a flying career by studying aviation in college and flight instructing. After gaining experience as a commuter airline pilot, she was selected for pilot employment at ValuJet in October of 1993. Upgraded to DC-9 captain in May, 1994, she had approximately 8,928 total flight hours, with 1,784 hours as pilot-in-command of the DC-9.

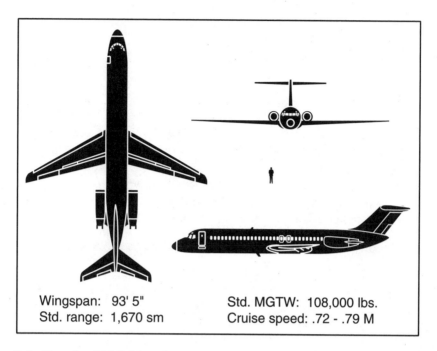

5-1 *Douglas DC-9-30 series.*

Wingspan: 93' 5"
Std. range: 1,670 sm
Std. MGTW: 108,000 lbs.
Cruise speed: .72 - .79 M

Seated to her right was First Officer Richard Hazen, 52. Retired from the U.S. Air Force, he had accumulated some 6,448 hours as a pilot, and an additional 5,400 hours as a civilian and military flight engineer. He had 2,148 hours in DC-9 aircraft, and was hired by ValuJet in November 1995.

About four minutes after takeoff Miami Departure Control instructed flight 592 to "turn left heading three zero zero, join the WINCO transition, climb and maintain one six thousand." First officer Hazen acknowledged that transmission.

Three minutes later, at 2:10 P.M., as the aircraft was climbing through 10,600 feet mean sea level (msl) at an indicated airspeed of 260 knots, an unidentified sound was heard in the cockpit. Captain Kubeck asked, "What was that?" followed by first officer Hazen's stating "I don't know."

"We've got some electrical problems," the captain said just twelve seconds later. "We're losing everything!"

At about the same time, unaware of the peril facing the ValuJet crew, the departure controller advised flight 592 to contact Miami Center. But serious problems were developing on the flight deck, and the captain urgently told the first officer, "We need, we need to go back to Miami!"

Three seconds later cries from the passenger cabin were heard, "fire, fire, fire, fire!" Shortly after that, a voice yelled, "We're on fire, we're on fire!"

Having not heard a reply to the frequency-change transmission, the departure controller reissued the instruction. The first officer quickly replied that they needed an immediate return to Miami. The controller replied, "Critter five ninety two, uh, roger, turn left heading two seven zero, descend and maintain seven thousand." The first officer acknowledged the heading and altitude.

"What kind of problem are you havin'?" asked the controller.

"Uh, smoke in the cockp...smoke in the cabin," radioed the first officer.

Aboard the DC-9, the situation was critical. The cockpit door opened, and a flight attendant pleaded, "OK, we need oxygen. We can't get oxygen back there."

At 2:11 P.M., the controller instructed flight 592 to turn left to a heading of 250 degrees, descend, and maintain 5,000 feet. At about this time the cockpit voice recorder (CVR) recorded additional shouting from the passenger cabin, followed by a flight attendant exclaiming, "completely on fire!"

The crew initiated a descent and a turn to the south. The first officer radioed, "Critter five ninety two, we need the uh, closest airport available."

The controller replied, "Critter five ninety two, they're gonna be standing [unintelligible], standing by for you. You can plan runway one two. When able, direct to Dolphin now."

The first officer radioed that they would require radar vectors to Dolphin, an electronic radio navigational aid located on the final approach to Runway 12. The controller then issued a vector of 140 degrees, and the first officer acknowledged the transmission.

A minute later the controller transmitted, "Critter five ninety two, keep the turn around heading, uh, one two zero." There was no response from the 592's crew. When the aircraft was descending through 7,200 feet on a heading of 218 degrees the flight data recorder (FDR) stopped recording data. Because the airplane's transponder continued to operate, airplane position and altitude data continued to be recorded by ATC radar.

At 2:13:18 P.M., the departure controller stated, "Critter five ninety two, you can, uh, turn left heading one zero zero and join the runway one two localizer at Miami." Again, there was no response. The controller again instructed flight 592 to descend and maintain 3,000 feet. A few seconds later, an unintelligible transmission was intermingled with a transmission from another airplane.

ValuJet 592 slammed into the Florida Everglades at a speed of over 400 miles per hour at 2:13:42 P.M., about seventeen miles northwest of the Miami International Airport. Occurring just ten minutes after takeoff, this tragedy claimed the lives of one hundred and ten persons and brought forth one of the most challenging and demanding aircraft accident investigations in history.

The investigation and findings

The National Transportation Safety Board was notified of the accident at about 2:30 P.M. That afternoon a full go-team of investigators assembled at the Federal Aviation Administration's (FAA) hangar at Washington National Airport, and launched for Miami in the FAA's Gulfstream G4 aircraft.

Investigators quickly learned from aerial and ground inspections that the aircraft had virtually disappeared into the Everglades. "It was if the ground sucked that plane up," said a witness who observed the impact as he fished nearby. At an evening news conference following their first full day on site, NTSB Vice Chairman Robert Francis explained that the day had been spent studying "the best way to deal with a very challenging, very difficult environment." *USA Today* described the environment that Mr. Francis referenced: "Black water coated with jet fuel. Saw grass with razor sharp edges. Dank muck that makes solid footing impossible. And alligators." Added to that list of challenges were mosquitoes, poisonous snakes, decomposing body parts, hot temperatures, water with visibilities that were described as ranging from "zero to none," and lingering concerns that spilled jet fuel may ignite.

The Florida Everglades are rich with vegetation, wildlife, and water. Sawgrass grows to be six or eight feet high. In the area of the crash site the water depth varied from about six inches to roughly eight to ten feet. At the bottom of this water was a thick layer of decomposing sawgrass root and thatch, all resting on a bed of limestone.

Only a crater in the mud and flattened sawgrass initially identified the DC-9's primary impact area. The centerline of the crater was oriented along a north/south axis, measuring 130 feet long and 40 feet wide. Most of the wreckage was located south of the crater in a fan-shaped pattern, with some pieces found more than 750 feet south of the crater.

During the seven weeks of the recovery effort, extensive precautionary measures were employed to protect salvage workers and divers from the hostile environmental elements. Sharpshooters were posted as sentries against alligators. Searchers had to don protective biohazard suits and fisherman's hip boots, a process that took up to an hour. Because of the extreme temperatures inside those suits, workers were limited to twenty to forty minutes of actual search time.

Due to drastically limited underwater visibility, workers and divers were forced to search blindly using their hands and nets. As wreckage was found it was placed on airboats (see Fig. 5-2) and transported to a nearby levee for decontamination, then carried by truck to a hangar for examination.

5-2 *Recovering the remains of ValuJet 592. . .* Courtsey FAA

The catastrophic impact and destruction of the aircraft, coupled with the difficult environmental conditions, precluded complete recovery of all aircraft components. But those parts that were recovered told a grim story.

The majority of both wings were found and identified. About fifty percent of the left aileron was located, as was about seventy-five percent of the right aileron. Actuators for the slats, flaps, and landing gear were found in their retracted positions, as were three of the wing spoiler panels. One spoiler was found at forty degrees deflection, but with evidence of impact damage. The right side of the forward fuselage was more severely fragmented than other areas, making specific part identification extremely difficult. Most of the smaller pieces were moved inside where they could be cleaned, sorted, and grouped with similar debris for further analysis (see Fig. 5-3).

Forward cargo bin reconstructed

Using cut-to-scale plywood and chicken wire to form a frame, investigators created a full scale, three-dimensional mockup of the forward cargo compartment. As pieces of wreckage were identified as being from that area, they were placed on the frame in their respective positions (see Fig. 5-4).

5-3 *Identifying and sorting the thousands of small pieces of the DC-9. . .* Courtsey FAA

5-4 *Reconstruction of the forward cargo compartment. . .* Courtsey FAA

Inspection of these components provided evidence of fire damage throughout the forward cargo compartment and areas directly above it, with the most severely fire damaged areas being the compartment's forward areas and ceiling. Inspection also revealed that other areas of the airplane did not exhibit significant fire damage, including the cockpit and the electronics compartment of the aircraft beneath the cockpit.

Recovered aircraft wiring was examined for evidence of arcing, fire, and heat damage, but none displayed short-circuiting or arcing that could have initiated the fire. The wire bundles that ran along the outside walls of the forward cargo compartment had burn patterns consistent with damage from an external source of heat.

The NTSB determined that the heat and fire damage to the interior of the forward cargo compartment was more severe than the damage to the exterior, consistent with a fire being initiated inside the cargo compartment. They reinforced this finding by noting that the cargo compartment liner, which was designed to keep a fire contained within the cargo compartment, would also have functioned to keep an externally initiated fire out of the compartment.

Flammable cargo

Interviews with ValuJet cargo handlers and a COMAT shipping ticket provided evidence that a COMAT shipment had been loaded in the forward cargo compartment of flight 592. This shipment consisted of five boxes of approximately 140 unexpended aircraft chemical oxygen generators, along with aircraft tire and wheel assemblies. The shipment was being sent to ValuJet's stores department in Atlanta from SabreTech, a maintenance organization in Miami that ValuJet had contracted to perform heavy maintenance. In preparing the shipment, a SabreTech stock clerk reported that he rearranged the oxygen generators in cardboard boxes and placed about two to three inches of plastic bubble wrap in the top of each box before closing and sealing them. Bubble wrap is a petroleum-based product, and can be highly flammable.

The ValuJet ramp agent who loaded the cargo in the forward cargo compartment told investigators, "I was stacking—stacking the boxes on the top of the tires." He testified that he remembered hearing a "clink" sound when he loaded one of the boxes, and that he could feel objects moving inside the box.

Chemical oxygen generators like those being shipped onboard flight 592 are normally used in aircraft passenger cabins to provide an emergency breathing source in the event of an aircraft decompression. When installed, they are mounted in panels above each row of passenger seats and are connected to oxygen masks (see Fig. 5-5). If

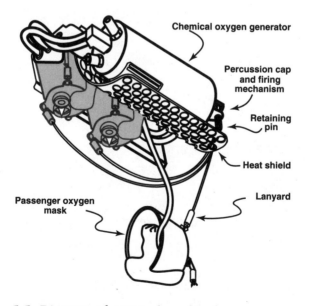

5-5 *Diagram of a typical overhead oxygen generator/mask installation.* . . . Courtsey NTSB

a rapid decompression occurs the panels will open and the masks will drop, making them available for passenger use. When the mask is pulled down, a spring-loaded initiation pin strikes a percussion cap containing a small explosive charge mounted in the end of the oxygen generator, starting a chemical reaction that produces breathable oxygen.

A by-product of this reaction, however, is the liberation of heat, which causes the exterior surface of the unit to reach 450 to 500 degrees Fahrenheit. At the time of the accident, the possible hazards associated with this rapid heat generation were well known. Ten years earlier a DC-10 was destroyed by fire while it sat on the ground at Chicago O'Hare International Airport. The cause of that mishap was attributed to inadvertent activation of a chemical oxygen generator that was being shipped in the cargo compartment. Conclusions from that investigation further revealed that the generators had not been packaged or labeled in accordance with federal regulations. Three days after that event the FAA notified all domestic air carriers and foreign airworthiness authorities of the circumstances surrounding the DC-10 fire and reminded operators that chemical oxygen generators were oxidizers and therefore classified as hazardous materials, which must be properly packaged and stowed securely.

Eighteen months later, a Boeing 757 experienced an in-flight cabin fire related to chemical oxygen generators. Flight attendants were only able to contain the fire after discharging three Halon fire extinguishers into the blaze. Since that incident, there were four additional known incidents involving oxygen generators, but catastrophes were averted in each case because the discovery of smoldering or flaming cargo did not occur in flight, but rather while loading or unloading cargo on the ground.

Outdated oxygen generators

In early 1996 ValuJet purchased two used McDonnell Douglas MD-82s (N802VV and N803VV) and one MD-83 (N830VV) from McDonnell Douglas Finance Corporation. All three aircraft, commonly referred to somewhat generically as "MD-80s," were ferried to the Miami maintenance and overhaul facility of SabreTech Corporation for various modifications and maintenance functions.

The chemical oxygen generators for the MD-80 had a manufacturer-imposed maximum allowable service life of twelve years from date of manufacture. Before ValuJet could place the MD-80s into service, they would have to ascertain that all life-limited components, including the on-board oxygen generators, were in compliance with the limitations. Because the majority of the generators on both MD-82s had exceeded this limit, ValuJet directed SabreTech to remove and replace all of them.

Contracting for vital services

Having outside contractors such as SabreTech perform work was not unusual for ValuJet. At the time of the accident, the airline had contracts with twenty-one FAA-certified maintenance facilities and repair stations to service its fifty-two airplanes. SabreTech was one of three such organizations that ValuJet used for heavy maintenance. ValuJet provided oversight of this work through on-site technical representatives; however, in many cases these individuals were contract employees themselves. These tech reps were responsible for ensuring that projects stayed on schedule and were carried out in accordance with contract specifications.

Like ValuJet, SabreTech also relied heavily on use of contract personnel. At the time the work was being done to remove and replace the oxygen generators from the ValuJet airplanes, six companies were the primary sources of SabreTech's contract workers. SabreTech records indicated that between February 1996 and May 1996, of the 587 individuals working on the three ValuJet MD-80 airplanes, 75 percent were employed by contract organizations, with the remainder being actual SabreTech employees.

Time pressure on the shop floor

According to the Aircraft Maintenance Services Agreement between SabreTech and ValuJet, SabreTech would be penalized $2,500 for each calendar day that an aircraft was delayed in delivery. Work on N802VV was to have been completed by April 24, 1996, and by April 30, 1996, for N803VV. Two days after work on N802VV was supposed to have been completed but was not, a SabreTech interoffice memorandum was issued, stating, "Effective immediately, due to the present workload schedule all Maintenance Personnel including Management are required to work seven days (including days off). We will return to regular work schedule when the three MD-80s are delivered." Mechanics later stated there was a great deal of pressure

to complete the work on the airplanes on time, including requirements that personnel work twelve-hour shifts, seven days per week.

Replacing and checking the passenger oxygen system was one of
the last items to be completed. For N802VV, this item was completed
on May 9, 1996—several days after the deadline and just two days
prior to the crash of flight 592. For N803VV, work on the passenger
oxygen system was completed on April 30, 1996, and the aircraft
was delivered to ValuJet on May 1, 1996—one day after the contract
delivery date.

Documentation for removing/replacing oxygen generators

A "routine work card" is a computer-generated, step-by-step listing
of the key procedures used to complete a particular maintenance
task. To indicate completion of these items, signatures are required
on the card after each step. SabreTech personnel used routine work
card 0069 to track the removal and installation of chemical oxygen
generators on customer aircraft. In the section specific to removal,
the first item on the card was a bold-print statement, which read:

> *"WARNING: UNEXPENDED OXYGEN GENERATORS CON-
> TAIN LIVE IGNITION TRAINS, AND, WHEN ACTIVATED,
> GENERATE CASE TEMPERATURES UP TO 500 DEGREES F.
> USE EXTREME CAUTION WHILE HANDLING TO PREVENT
> INADVERTENT REMOVAL OF FIRING PIN. IF GENERATOR
> SHOULD BECOME ACTIVATED, IMMEDIATELY PLACE ON A
> NONCOMBUSTIBLE SURFACE."*

Following that warning was a seven-step procedure for removing
the generator. The second step stated, "If generator has not been expended, install shipping cap on firing pin." A shipping cap, also
known as a safety cap, is a plastic cover placed over the percussion
cap to prevent accidental activation of the chemical oxygen generator during shipping and handling.

Work card 0069 referred to Douglas Maintenance Manual (MM)
Chapter 35-22-03, titled, "Passenger oxygen insert units—maintenance practices." With minor differences, this chapter contained es-
sentially the same information as work card 0069. Of significance is
that neither of these documents referred to another relevant chapter
of the MM, Chapter 35-22-01, titled "Chemical oxygen generator—
maintenance practices."

Unlike the seven-step procedure detailed by work card 0069, Chapter 35-22-01 had an eight-step procedure for oxygen generator removal. The first seven steps corresponded to those on work card 0069, but one additional step stated: "Store or dispose of oxygen generator (Ref. paragraph 2.C. or 2.D.)."

Paragraph 2.C stated:

"(1) Oxygen generators must be stored in safe environment.

(2) Each unit shall be checked before placing it in storage to assure that release pin restraining firing mechanism is correctly installed.

(3) All serviceable and unserviceable (unexpended) oxygen generators (canisters) are to be stored in an area that ensures that each unit is not exposed to high temperatures or possible damage."

Paragraph 2.D stated, "No oxygen generator (canister) is to be disposed of until it is initiated and chemical core is fully depleted." It contained a five-step procedure for purposely expending the generator.

In their investigation, the NTSB commented, "Although work card 0069 warned about the high temperatures produced by an activated generator, it did not mention that unexpended generators required special handling for storage or disposal, that out-of-date generators should be expended and then disposed of, or that the generators contained hazardous substances/waste even after being expended; further, the work card was not required to contain such information. Although these issues are addressed in the Douglas MD-80 Maintenance Manual, Chapter 35-22-01...the mechanics likely completed the removal of the generators without referring to this section of the maintenance manual."

The NTSB said, "The Safety Board concludes that had work card 0069 required and included instructions for expending and disposing of the generators in accordance with the procedures in the Douglas MD-80 maintenance manual, or referred to applicable sections of the maintenance manual, it is more likely that the mechanics would have followed at least the instructions for expending the generators."

The NTSB also expressed concern over SabreTech's lack of record keeping for maintenance tasks. They noted that seventy-two individuals logged approximately 910 hours against the tasks described on work card 0069. However, there were no records to show who did which tasks, other than the signature of the mechanic who ultimately signed the work cards signifying completion.

Although the NTSB's report did not address it, some NTSB officials privately wondered if a language barrier might have been a factor. While all the SabreTech manuals and work cards were printed in English, the aviation standard, estimates placed the number of SabreTech workers who spoke foreign languages as their "first language" at 70 percent.

Removal of oxygen generators: The maintenance process

In mid-March 1996, SabreTech crews began removing oxygen generators from the MD-82, and replacing them with new units. As did all oxygen generators manufactured by Scott Aviation since 1988, these new generators each had a label that stated:

"WARNING—THIS UNIT GETS HOT! WHEN REMOVING THIS UNIT INSTALL SAFETY CAP OVER PRIMER—DO NOT PULL LANYARD

IF ACTIVATED PLACE ON SURFACE THAT WON'T BURN"

According to the SabreTech mechanic who signed work card 0069 for N802VV, when an old oxygen generator was removed from the MD-82s, a green SabreTech "Repairable" tag was attached to the body of the canister. Another mechanic stated that he ran out of green tags and attached a white "Removed/Installed" tag on four to six generators. According to SabreTech's FAA-approved inspection procedures manual, because these generators were not repairable items, a red "Condemned Parts" tag should have been used instead of green or white tags. In the "reasons for removal" section of the tag mechanics wrote various entries, such as "out of date," "expired," and "outdated," to signify that the generators had been removed because of a time limit being exceeded. After being tagged the generators were placed in a cardboard box and placed on a rack near the airplane.

The mechanic who signed work card 0069 for N802VV told investigators that he was aware of the need for safety caps. He stated that some mechanics had discussed using caps from the new generators, but they rejected that idea because those caps had to remain on the new generators until the final "mask drop check" was completed at the end of the process. If the caps were removed prior to the mask drop check, there would be the distinct possibility that some generators could activate when the masks were dropped. In a post-accident interview, a mechanic stated that a SabreTech supervisor told him that the company did not have any additional safety caps available. However, when questioned by NTSB officials, the supervisor denied that anyone had ever mentioned to him the need for safety caps.

According to the SabreTech director of logistics, the Miami facility had never before performed this work, so safety caps were not carried in SabreTech's parts inventory. SabreTech considered these items as "peculiar expendables," which as defined in the Aircraft Maintenance Services Agreement between SabreTech and ValuJet, meant that ValuJet had final responsibility to supply them.

Packaging and shipping of oxygen generators

By the first week in May 1996, most of the old oxygen generators had been collected in five cardboard boxes. These boxes were then moved to the shipping and receiving area and placed on the floor near shelves storing other ValuJet parts.

Earlier customer audits had revealed "housekeeping" problems in the shipping/receiving area, so when faced with a visit by a potential customer, the SabreTech director of logistics directed employees to clean up the space and to remove all of the items on the floor. In a post-accident interview he stated that when he gave the edict to move the boxes, he did not know their contents, nor did he give specific instructions concerning their disposition.

That same day the director of logistics spoke with one of the ValuJet technical representatives concerning disposition of these parts. According to the maintenance services contract, it was SabreTech's responsibility to store excess ValuJet parts until accepted by the airline; however, the airline decided that it would be at least May 13th before they would accept the parts.

On May 8 a SabreTech stock clerk asked the director of logistics, "How about if I close up these boxes and prepare them for shipment to Atlanta?" When the director responded, "Okay, that sounds good to me," the clerk redistributed the oxygen generators in the boxes, placing them end-to-end along the length of the box. He placed two to three inches of plastic bubble wrap in the top of each box, closed and sealed them, and applied a ValuJet COMAT label on each box with the notation "aircraft parts."

The next day the clerk asked a receiving clerk to prepare a shipping ticket for the five boxes and three DC-9 tires, instructing him to write, "Oxygen Canisters—Empty" on the shipping ticket. In preparing the ticket, the receiving clerk shortened the word "Oxygen" to "Oxy" and then put quotation marks around the word "empty."

The shipping clerk stated that he identified the generators as "empty canisters" because none of the mechanics had talked to him about what they were or what state they were in, and that he had just found the boxes on the floor one morning. He stated that he did not know what the items were, and when he saw green tags on them, he assumed that meant they were empty.

On May 10 the stock clerk asked a SabreTech driver to deliver the items listed on the COMAT shipping ticket to ValuJet, but the driver was busy so the shipment was not picked up until May 11. The driver transported the items to the ValuJet ramp and at the request of a ValuJet ramp agent, loaded the material onto a baggage cart. The ValuJet employee signed the shipping ticket and the driver returned to the SabreTech facility.

Oxygen generator fire tests

At the request of the NTSB, in November 1996 a series of fire tests involving chemical oxygen generators were conducted at the FAA's fire test facility at Atlantic City, New Jersey. Their purpose was to provide data regarding the overall nature of a fire initiated by an oxygen generator and fed with high concentrations of oxygen released from additional oxygen generators. Each of the five tests was carried out in an instrumented and fire-protected test chamber. To simulate conditions on flight 592, three of the tests utilized five cardboard boxes, each containing twenty-four generators. Two inches of plastic bubble wrap was placed in the top of each box, and the boxes were sealed. To start the fire, a retaining pin on one of the generators located at the top of a box was pulled. Thermocouples and a water-cooled calorimeter were located 40 inches above the chamber floor, which is the distance between the floor and ceiling of a DC-9 cargo compartment.

Two of the tests produced only minor smoke generation. However, the remaining three tests had dramatic results. In each of those tests, within as little as ten minutes, temperatures in the test chamber reached more than 2,000 degrees F.

In the final test, two boxes of oxygen generators were placed on top of a main landing gear tire pressurized to 50 psi. (The NSTB consulted with Goodyear and learned that damage to one of the shipped main tires was consistent with a tire rupturing under pressure of between 30 and 50 psi.) The remaining three boxes of oxygen generators were placed around the tire and surrounded by luggage. About ten minutes after the retainer pin was pulled, the temperature reached 2,000 degrees F; after eleven minutes, the temperature was about 2,800 degrees F. About eleven-and-a-half minutes after ignition the temperature exceeded the measurement capabilities of the instrument (3,200 degrees F). Sixteen minutes after ignition the tire ruptured.

Conclusions

After fifteen months of intense investigation the NTSB concluded their investigation. They found no preexisting aircraft mechanical malfunctions that could have contributed to the accident, nor was weather a contributing factor.

Wreckage recovery revealed sufficient evidence of fire damage throughout the forward cargo compartment and areas directly above it, while other areas exhibited no significant fire damage. There was no evidence of electrical short circuits or arcing that could have initiated a fire.

The Board's investigation focused on the COMAT shipment that had been loaded into the forward cargo compartment. They considered the fact that safety caps were not installed over the percussion caps of the oxygen generators, and the generators were packaged in a manner that would allow them to bump around during handling. They also considered that the severity of the fire damage indicated an extremely high degree of heat exposure.

The NTSB stated, "Based on the results of the Safety Board's fire tests on chemical oxygen generators that were conducted near Atlantic City, New Jersey, after the accident, the physical evidence of

fire damage in the forward cargo compartment of the accident airplane, and the lack of other cargo capable of initiating a fire in the
forward cargo hold, the Safety Board concludes that the activation
of one of more chemical oxygen generators in the forward cargo
compartment of the airplane initiated the fire on ValuJet flight
592." The NTSB found that the failure to install the shipping caps
was a critical error because had they been properly installed, the
generators would not have activated and the accident would not
have occurred.

The Safety Board's analysis and recommendations focused on the
initiation and propagation of an onboard fire and efforts that can be
employed by the FAA and industry to mitigate hazards of cargo compartment fires.

Fire initiation and propagation

The first indication of a problem occurred at 2:10:03 P.M., when the
CVR recorded an unidentified sound. Simultaneously, the FDR
recorded anomalies in altitude and airspeed parameters. Recorded
indicated altitude dropped 817 feet and recorded indicated airspeed
decreased by thirty-three knots. Interestingly, however, the values
returned to their normal values in less than four seconds. The Safety
Board deduced that these FDR changes were consistent with a static
pressure increase of about sixty-nine pounds per square foot inside
the aircraft.

In the Safety Board's fire tests, a main gear tire that was inflated to
fifty psi ruptured sixteen minutes after an oxygen generator was activated. Since the tires in the cargo compartment of ValuJet 592 were
located just above the left static ports, and those ports are the supply data for FDR altitude and airspeed measurements, investigators
concluded that the unidentified sound on the CVR and the FDR
anomalies at 2:10:03 P.M. were most likely caused by the rupturing
of the tire after being heat damaged. (For more information on static ports, see Chapter 4, "A tale of two tragedies.")

The Safety Board attempted to determine exactly when the oxygen
generators may have activated. One theory was they were activated
when the boxes were jostled while being loaded into the cargo compartment. However, the cargo loading process was complete by 1:40
P.M., and the tire did not rupture until 2:10 P.M., thirty minutes later.
Alternatively, since the boxes were not secured with cargo restraints,

it was suggested that they could have shifted during takeoff. This, however, was only six to seven minutes before the tire ruptured.

In any case, the Safety Board realized that many factors could alter the rate at which a fire might propagate. They concluded that "one or more of the oxygen generators likely were actuated at some point after the loading process began, but possibly as late as during the airplane's takeoff roll."

Flight profile after fire eruption

At 2:10:15 P.M., twelve seconds after the apparent sound of the tire rupturing was recorded on the CVR, the captain commented on electrical problems. The NTSB determined that these electrical complications likely resulted from insulation burning on wires that ran outside of, and adjacent to, the forward cargo compartment. At this point investigators believed that the fire had escaped the cargo compartment, but it had not yet penetrated into the passenger cabin, which likely did not occur until 2:10:25 P.M., when shouts were heard from the cabin, stating, "fire, fire, fire, fire!"

Just then, the FDR recorded the right engine parameters indicating idle thrust, but the left engine remained at its previous value. About this time the CVR recorded the sound of the landing gear warning horn, which activates when one or both engine thrust levers are retarded in flight and the landing gear is not down. This is a likely indication that the crew was attempting to retard the thrust to initiate an immediate descent, but able to do so only on one engine. "Because the flightcrew would not have intentionally reduced thrust on one engine only, they must have been unable to reduce thrust on the left engine because of fire damage to the engine control cable located under the compartment," the NTSB stated.

At 2:12:00 P.M., the altitudes recorded on the FDR decreased, but they no longer corresponded with those being recorded by ATC radar. It was determined that this was further indication that the fire damage was continuing to increase.

Radar data indicated that at about 2:13 P.M., the DC-9 began a steep left turn toward the Miami International Airport and began a rapid descent out of 7,400 feet. For the next thirty-two seconds the descent rate averaged 12,000 feet per minute (fpm), and the airplane turned from a southwesterly heading to an easterly heading (see Fig. 5-6).

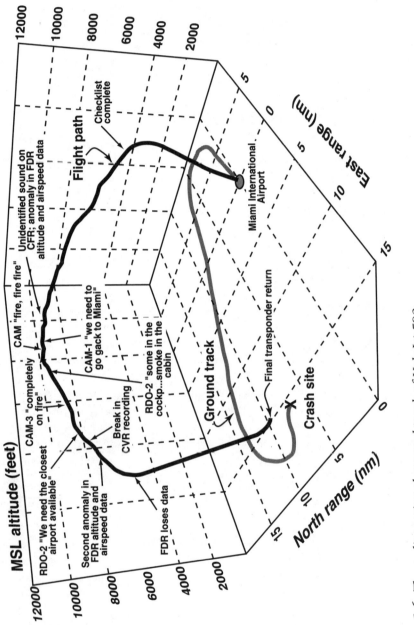

5-6 *Three-dimensional ground track of ValuJet 592.* . . Source: NTSB

An aircraft with asymmetric thrust (one engine at a higher power setting than the other) will tend to turn in the direction corresponding to the engine with lower thrust. In flight 592's case, the thrust asymmetry would have tended to turn the aircraft to the right unless counteracted by the crew. The NTSB highlighted that flight 592 turned to the left and stopped on a heading that placed the Miami International Airport directly in front of the aircraft.

Further, when the last ATC transponder radar data from flight 592 was recorded at 900 feet, the rapid descent was being reduced. "The control inputs required to balance asymmetric thrust during the steep left turn...and level-off, indicates that the...captain and/or first officer were conscious and applying control inputs to stop the steep left turn and descent (until near 2:13:34)," the NTSB determined. "Thus, the airplane remained under at least partial control by the flightcrew for about three minutes and nine seconds after 2:10:25."

Because of the lack of evidence from the crash site and because data was missing from both the CVR and FDR, the Safety Board was unable to positively ascertain why control was lost shortly before ground impact. However, wreckage examination provided clear indications that sometime in flight the left side floor beams had melted and collapsed, likely jamming the control cables on the captain's side. The Board commented that while the first officer might have been able to take over flying duties at this point, it was possible that his control cables were also compromised. The Safety Board concluded that the most likely cause of the loss of control was due to flight control failure as a result of extreme heat and structural collapse. However, they did not rule out that the flightcrew may have become incapacitated in the last seven seconds of flight.

The NTSB remarked that there were no indications that the flightcrew donned their crew oxygen masks. The Board recognized that putting on oxygen masks and smoke goggles can be time consuming and distracting, but they emphasized the importance of donning them at the first indication of smoke anywhere in the aircraft. They also urged the industry to take advantage of emerging technologies to develop oxygen masks and goggles that could be used more easily by crews.

Cargo compartment fire/smoke detection and extinguishing

The forward cargo compartment of the DC-9 was classified as a "Class D" cargo compartment and therefore was not equipped (or required to be equipped) with smoke detection or fire suppression capability. The NTSB observed that the crew, therefore, would have no means of detecting the fire until its smoke and flames had reached the passenger cabin, nor would they have had the means to suppress it once it was manifested.

The NTSB noted that in response to an in-flight fire they had investigated several years earlier, a safety recommendation to the FAA was issued to require fire/smoke detection and fire extinguishing systems for all class D cargo compartments. The FAA did not implement that recommendation after conducting the government-required cost/benefit analysis and determining that the costs of such a system would outweigh anticipated benefits. As it pertained to the ValuJet accident, the NTSB stated, "The Safety Board concludes that had the FAA required fire/smoke detection and fire extinguishment systems in Class D cargo compartments, as the Safety Board recommended in 1988, ValuJet flight 592 would likely not have crashed. Therefore, the failure of the FAA to require such systems was causal to this accident."

The way ValuJet conducted business

When four investors each provided $1 million to form ValuJet, company president Lewis Jordan stated "We're interested in being the Wal-Mart of the airline business." His desire was to be the industry leader in high-volume, lost-cost operations. ValuJet began operations in October 1993 with a fleet of two aircraft and serving three cities, but by the time of the accident two-and-a-half years later, the Atlanta-based airline had expanded to fifty-two aircraft, serving thirty-one cities. Their fleet consisted of three types of DC-9 aircraft, including MD-80s. Because they were obtaining used aircraft from several different sources, there were eleven different configurations of aircraft.

During the investigation the NTSB interviewed ValuJet's former senior vice president of operations, who had been responsible for both flight operations and maintenance until his retirement in February 1996. He stated that when he joined ValuJet in June 1994, he found a number of discrepancies, including maintenance records that he described as "not in great shape." He also stated that he found "lots of sloppiness due to rapid growth" and that the person who was serving as director of quality assurance in the maintenance department was in "way over his head."

This individual stated that in 1994 the airline was adding airplanes "as fast as we could get them" and his concern over this widespread growth caused him to discuss the matter with the company's president. Because of that conversation, he and the president decided to limit the rate of expansion to eighteen to twenty-one airplanes per year.

In June 1994, ValuJet made a public stock offering. By the end of the first day of public trading the stock price had increased twenty-nine percent, and by November 1995 the stock had soared to more than 400 percent of its initial offering price. "We're the great American success story," Mr. Jordan said, "the darling of the airline industry, the darling of Wall Street."

Around the time of the accident, ValuJet was being unofficially referred to by some in the aviation industry as a "virtual airline." Most aircraft maintenance and servicing had been "outsourced," as had some aspects of flight operations. FlightSafety International had been contracted to interview, hire, and train all ValuJet pilots. But only after a candidate had paid FlightSafety the $9,500 cost of his or her flight training and successfully completed it, would ValuJet officially hire them.

The NTSB recognized that air carriers can successfully contract many operational functions, but also remarked, "To properly oversee a subcontractor, the air carrier must verify that the subcontractor has in place the necessary equipment and procedures to perform the air carrier's work, that the individuals hired by the subcontractor to perform the work are qualified and capable of performing it, and that those individuals are in fact using proper equipment and procedures." The Board was highly critical of ValuJet's on-site quality assurance safeguards, particularly in dealings with SabreTech and other maintenance contractors. They concluded, "ValuJet failed to adequately oversee SabreTech and this failure was a cause of the accident."

FAA oversight

Previously, the FAA had conducted a regional and national inspection of ValuJet and had uncovered several deficiencies in their flight operations and in-house maintenance functions. The NTSB stated that the FAA Atlanta Flight Standards District Office (FSDO) reacted properly in February 1996, when they targeted ValuJet for more intense surveillance, and by precluding future ValuJet expansion until these problems were rectified.

On February 29, 1996, in correspondence with Lewis Jordan, the Atlanta FSDO wrote, "ValuJet Airlines has recently experienced four occurrences." The letter continued, "These occurrences, coupled with the preliminary findings of the Federal Aviation Administration's Special Emphasis Review completed on February 28, 1996, give us concern that ValuJet is not meeting its duty to provide services with the highest possible degree of safety in the public interest...It appears that ValuJet does not have a structure in place to handle your rapid growth, and that you may have an organizational culture that is in conflict with operating to the highest possible degree of safety."

The "four occurrences" referenced in the FAA's letter included a DC-9 landing in 1995 with only fourteen minutes of fuel remaining; a gear retraction problem the next year where the crew continued to the destination rather than returning, leading to an accident upon landing; and two cases of aircraft running off the runway upon landing.

In spite of these efforts by the FAA, the NTSB felt that it was too little, too late. "[T]he Safety Board is concerned about the timelines of this action," stated the NTSB. "By the time ValuJet's growth was halted, it had already outgrown its capability to adequately coordinate and oversee its maintenance functions. This should have been apparent to the FAA earlier, especially given the exceptional pace at which ValuJet was adding airplanes and routes, and its continued outsourcing of its heavy maintenance functions."

The NTSB stated that although the FAA inspections were comprehensive, they failed to "take into account the extent to which ValuJet had contracted out its operations and maintenance functions." The Safety Board remarked that the Principal Maintenance Inspector (PMI) never completed an inspection of SabreTech, although they were conducting a large portion of ValuJet's heavy maintenance. The Safety Board concluded that "the FAA's inadequate oversight of ValuJet's maintenance functions, including its failure to address ValuJet's limited oversight capabilities, contributed to this accident."

The FAA was likely anticipating that blow from the NTSB. Just weeks after the accident, but well before the NTSB's hearing, then-FAA administrator David Hinson testified to Congress. "We should have better understood the effects of rapid growth on this airline," said Mr. Hinson. He stated that ValuJet's growth "created problems that should have been more clearly recognized and dealt with sooner and more aggressively."

The probable causes

When the NTSB held their final hearing on ValuJet 592, intense public scrutiny required the use of a large hotel ballroom to accommodate the more than 400 family members of victims, members of media, and the general public. By the end of the day the NTSB had formed an opinion.

"The National Transportation Safety Board determines that the probable causes of the accident, which resulted from a fire in the airplane's class D cargo compartment that was initiated by the actuation of one or more oxygen generators being improperly carried as cargo, were (1) the failure of SabreTech to properly prepare, package, and identify unexpected chemical oxygen generators before presenting them to ValuJet for carriage; (2) the failure of ValuJet to properly oversee its contract maintenance program to ensure compliance with maintenance, maintenance training, and hazardous materials requirements and practices; and (3) the failure of the Federal Aviation Administration (FAA) to require smoke detection and fire suppression systems in class D cargo compartments."

"Contributing to the accident was the failure of the FAA to adequately monitor ValuJet's heavy maintenance programs and responsibilities, including ValuJet's oversight of its contractors, and SabreTech's repair station certificate; the failure of the FAA to adequately respond to prior chemical oxygen generator fires with programs to address the potential hazards; and ValuJet's failure to ensure that both ValuJet and contract maintenance facility employees were aware of the carrier's 'no-carry' hazardous materials policy and had received appropriate hazardous materials training."

In closing the hearing, NTSB chairman Jim Hall made a personal observation: "The ValuJet accident resulted from failures all up and down the line."

Recommendations

In concluding this investigation, the NTSB issued the following recommendations to the FAA:

- Expedite final rule-making to require smoke detection and fire suppression systems for all class D cargo compartments. (A-97-56)

- Specify that air carrier aircraft must have an operational cockpit-cabin interphone before being dispatched, and when this system is inoperative, the minimum equipment list (MEL) should not allow for dispatch. (A-97-57)

- Issue guidance to air carrier pilots about the need to don oxygen masks and smoke goggles at the first indication of a possible in-flight smoke or fire emergency and require that smoke goggles be packaged in such a way that they can be easily opened by the flightcrew. (A-97-58, A-97-60)

- Establish performance standards for the rapid donning of smoke goggles and ensure that pilots meet this standard through improved smoke goggle equipment and/or training. (A-97-59)

- Evaluate cockpit emergency vision technology and take action, as appropriate. (A-97-61)

- Evaluate and support appropriate research to develop technologies for enhancing passenger respiratory protection from toxic atmospheres resulting from in-flight and post-crash fires involving transport-category airplanes. (A-97-62)

- Evaluate the effectiveness of current DC-9 and other transport-category aircraft procedures for cabin smoke and fume evacuation. (A-97-63)

- Require airplane manufacturers to amend company maintenance manuals for airplanes that use chemical oxygen generators to indicate that generators that have exceeded their service life should not be transported unless they have been actuated and their oxidizer core has been depleted, and that warning labels be affixed to these oxygen generators to effectively communicate the dangers posed by unexpended generators and that they be considered hazardous materials. (A-97-64, A-97-66)

- Require that routine work cards used during maintenance of Part 121 aircraft (a) provide instructions for disposal of any hazardous materials or a direct reference to the maintenance manual provision containing those instructions and (b) include an inspector's signature block on any work card that calls for handling a component containing hazardous materials. (A-97-65)

- Require all air carriers to develop and implement programs to ensure that aircraft components that are hazardous (other than chemical oxygen generators) are properly identified and that effective procedures are established to safely handle those components after they are removed from aircraft. (A-97-67)

- Evaluate and enhance FAA oversight techniques to more effectively identify and address improper maintenance activities, especially false entries. (A-97-68)

- Review current industry practices and, if warranted, require that all maintenance facilities ensure that items delivered to shipping and receiving and stores areas of the facility are properly identified and classified as hazardous or nonhazardous, and that procedures for tracking the handling and disposition of hazardous materials are in place. (A-97-69)

- Include, in its development and approval of air carrier maintenance procedures and programs, explicit consideration of human factors issues, including training, procedures development, redundancy, supervision, and the work environment, to improve the performance of personnel and their adherence to procedures. (A-97-70)

- Review the issue of personnel fatigue in aviation maintenance; then establish appropriate duty time limitations for maintenance personnel. (A-97-71)

- Issue guidance to air carriers on procedures for transporting hazardous aircraft components, then require principal operations inspectors to review and amend, as necessary, air carrier manuals to ensure that air carrier procedures are consistent with this guidance. (A-97-72)

- Require air carriers to ensure that all maintenance facility personnel are provided initial and recurrent training in hazardous materials recognition, and in proper labeling, packaging, and shipping procedures. (A-97-73)

- Ensure that all air carriers' maintenance activities receive the same level of FAA surveillance, regardless of whether those functions are performed in house or by a contract maintenance facility. (A-97-74)

- Review the volume and nature of the work requirements of principal maintenance inspectors assigned to 14 CFR Part 145 repair stations that perform maintenance for Part 121 air carriers, and ensure that these inspectors have adequate time and resources to perform surveillance. (A-97-75)

- Develop programs to educate passengers, shippers and postal customers about the dangers of transporting undeclared hazardous materials aboard aircraft and about the need to properly identify and package hazardous materials before offering them for air transportation. (A-97-76, A-97-79, A-97-82)
- Instruct principal operations inspectors to review their air carriers' procedures for manifesting passengers, including lap children, and ensure that those procedures result in a retrievable record of each passenger's name. (A-97-77)

The NTSB also recommended that the U.S. Postal Service develop a program to help Postal Service employees identify undeclared hazardous materials being offered for transportation, and that they seek further civil enforcement authority when undeclared hazardous materials are identified. (A-97-80, A-97-81)

Industry action

One of the first responses to the accident was the issuance by the FAA of a Notice of Proposed Rule-Making (NPRM) in June of 1997 which would require the U.S. airline industry to within three years, install fire/smoke detection and fire extinguishment systems in Class D cargo compartments. The FAA issued that final rule, which upgraded all current class D cargo compartments on February 12, 1998. Those compartments must meet the same smoke detection and suppression standards as applicable to class C compartments by the year 2001, and no future aircraft may be designed with cargo compartments that do not meet the improved standard.

After a thorough review of PA and crew interphone system failure data, the FAA determined that adequate written procedures are not always available to the flightcrew. Accordingly, on July 24, 1998, Policy letter #9 to the Master Minimum Equipment List (MMEL) for all transport aircraft was issued, establishing operational standards for dispatching with maintenance issues affecting those systems.

In support of the Board's recommendation regarding early use of protective breathing equipment, the FAA issued a Flight Standards Information Bulletin (FSIB) directing Principal Operations Inspectors (POIs) to ensure that "each holder's approved training program reflects sufficient hands-on training for protective breathing equipment...." Furthermore, that program must have a requirement for "...checking each item of protective breathing equipment...[and] guidance to flight crewmembers to don both smoke goggles and oxygen masks at the first indication of any unidentified odor...[and] within the fifteen-second design objective...." (for additional information, see Chapter 6, A pilot's nightmare).

The FAA has evaluated all known cockpit emergency vision technology and has approved the installation of one type of device. But they determined that it may not be "necessary or appropriate to mandate the installation of this [or any other] system."

In reviewing all of the various techniques employed by various manufacturers for in-flight aircraft smoke evacuation, the FAA believes any specific procedure should remain an option to the operator rather than a mandatory requirement. The FAA's primary regulatory effort, therefore, will remain in the areas of fire prevention and suppression.

On December 30, 1996, the Research and Special Programs Administration (RSPA) issued a final rule prohibiting the transportation of all oxygen generators, expended or not, as cargo aboard passenger-carrying aircraft. Aircraft manufacturers have revised their maintenance manuals to reflect the new regulations.

As an immediate response to the Safety Board's concerns regarding maintenance work cards, the FAA issued Flight Standards Handbook Bulletin (FSHB) 98-10 to assist all maintenance personnel in:

- Recognizing aircraft components that contain hazardous materials.
- Utilizing proper storage, handling, packaging, and disposal procedures.
- Identifying specific hazards associated with those materials.

The Airworthiness Inspector's Handbook and AC 120-16D, Continuous Airworthiness Maintenance Programs, will also be revised to incorporate this information.

The FAA has initiated a study to "define and specify human factors issues in developing and approving air carrier and repair station maintenance procedures and programs." Once completed, regulatory action or policy change will be implemented.

An industrywide HAZMAT/COMAT conference was held in Washington, D.C., in November of 1996, which focused on many of those areas highlighted by the Board's recommendations. One month later the RSPA published an Advisory Notice, "Transportation of COMAT by Aircraft." In February of 1997 the same agency issued a Safety Alert, "Advisory Guidance: Offering, Accepting and Transporting Hazardous Materials." In early 1998, the FAA and RSPA copublished yet another safety alert on the special requirements of oxygen generators, and a brochure identifying specific information about dangerous COMAT was distributed by the FAA later that year.

In December of 1997, a FSHB was issued to assure proper surveillance of all air carrier maintenance programs. The adequacy of the organization and competency of maintenance personnel are to be continually checked, "whether those functions are performed in-house or by a contract maintenance facility."

Finally, on February 13, 1998, the FAA issued another FSHB, 98-04, directing all POIs to ensure that their operators comply with existing regulations regarding preparation of load and passenger manifests. Specifically, the term "passenger" applies to any passenger on board, regardless of age.

Epilogue

The Everglades tragedy brought an incredible amount of attention to ValuJet, SabreTech, and the organization responsible for providing oversight to those companies and the entire U.S. aviation industry—the FAA.

On June 17, 1996, just six weeks after the crash of flight 592, ValuJet agreed to suspend all revenue flight operations and surrender its operating certificate to the FAA. In signing a consent order, ValuJet agreed to pay the FAA $2 million as remedial, not punitive, payment "for costs incurred by the FAA to investigate, review, establish, reinspect, and ultimately enforce" the consent order.

That same day, a top-ranking FAA official, Anthony Broderick, associate administrator for regulation and certification, announced his retirement.

ValuJet spent the summer of 1996 operationally regrouping, until it could satisfy the FAA that its processes, practices, and procedures were sound. On September 30, 1996, ValuJet once again opened its doors to revenue passengers. On July 10, 1997, ValuJet announced plans to merge with AirTran Airways, and assume that company's name and identity.

On January 15, 1997, SabreTech voluntarily ceased operations at its Miami repair station, and surrendered its Orlando facility license to the FAA.

For the first time in U.S. history, criminal charges were filed as a result of a civilian, domestic airline accident. On July 13, 1999, the state of Florida and a federal grand jury indicted SabreTech, Inc., two of its mechanics, and a maintenance director. Charges included conspiracy to cover up problems that led to the crash and failure to properly train employees in the handling of hazardous materials. The state also brought 110 counts of third-degree murder, 110 counts of manslaughter, and one count of unlawful transportation of hazardous waste.

References and additional reading

Evans, David. 1997. Fly the fiery skies. *Washington Monthly*. September, 44.

Flight Safety Foundation. 1997. Chemical oxygen generator activates in cargo compartment of DC-9, causes intense fire and results in collision with terrain. *Accident Prevention*. November, 1-24.

Garrison, Peter. 1998. Final ValuJet report. *Flying*. April, 99.

McKenna, James T. 1997. Chain of errors downed ValuJet. *Aviation Week & Space Technology*. 25 August, 34-35.

National Transportation Safety Board. 1996. *Aircraft accident report: Uncontained engine failure/fire. ValuJet Airlines flight 597. DC-9-32, N908VJ. Atlanta, Georgia. June 8, 1995. NTSB/AAR-96/03.* Washington, D.C.: NTSB.

National Transportation Safety Board. 1997. *Aircraft accident report: Inflight activation of ground spoilers and hard landing. ValuJet Airlines flight 558. DC-9-32, N922VV. Nashville, Tennessee. January 7, 1996. NTSB/AAR-96/07.* Washington, D.C.: NTSB.

National Transportation Safety Board. 1997. *Aircraft accident report: Inflight fire and impact with terrain. ValuJet Airlines flight 592. DC-9-32, N904VJ. Everglades, Near Miami, Florida. May 11, 1996. NTSB/AAR-97/06.* Washington, D.C.: NTSB.

Nelms, Douglas. 1998. Where there's smoke... *Air Transport World*. June, 80-83.

Twigy, Bob, and Castaneda, Carol. 1996. Plane had history of problems. *USA Today*. 13 May, 1.

Twigy, Bob, Castaneda, Carol, and Sharp, Deborah. 1996. Alligators, access, saw grass and snakes will hamper effort. *USA Today*. 13 May, 3.

6

A pilot's nightmare

The in-flight fire of FedEx 1406

Operator: Federal Express

Aircraft Type: Douglas DC-10-10

Location: Newburgh, New York

Date: September 5, 1996

Every day, thousands of tons of cargo are flown on commercial air-craft, linking businesses in cities large and small throughout the world. In just the last fifteen years, several "small package" airlines have established intricate delivery networks to support their rapidly growing clientele. Subject to most of the same operational rules and requirements as major passenger airlines, these operators provide an efficient, safe, and financially successful freight transportation service.

Occasionally, hazardous cargoes such as corrosive or radioactive material, compressed gasses, or flammable liquids are accepted for shipment by cargo-only airlines. Usually they are properly labeled, packaged, and sealed, and thus can be safely transported, but some-times they are not. The greatest risk posed by these materials (see Chapter 5, "Carnage in the Everglades") is the one most dreaded by all aviators: an in-flight fire.

Flight history and background

FedEx flight 1406 was scheduled to depart Memphis at 2:42 A.M. (CDT) on September 5, 1996. Eighty-two thousand pounds of cargo were loaded into thirty-six cargo containers and one open pallet on the DC-10 (see Fig. 6-1) for the first leg of the Boston round-trip "all-nighter." Originally acquired by Continental Airlines in 1975, this DC-10 entered service with FedEx in 1985.

Fight planning and crew briefings proceeded normally, including a routine discussion with one of the airline's "dangerous goods" experts. This specialist verified the location of all known hazardous cargo on the aircraft as being loaded in containers 1L/1C and 3R (as shown in Fig. 6-2), near the very front of the cabin cargo area. He further advised the flightcrew that Halon (an inert chloroflourocarbon gas that is used as a fire extinguishing agent) hose connections had been established with the only designated dangerous goods container, 1L/C. The captain of flight 1406 then received the Notification of Dangerous Goods Loading Form, documenting the proper shipping name, hazard class, location aboard the aircraft, and other information about each dangerous article being shipped. He signed the document and kept the "Part A" envelope, containing copies of all the shipping forms.

Another document, Notification of Loading of Dangerous Goods (Part B or C) was affixed directly to each package and contained very specific information regarding the exact contents of the shipment and a twenty-four-hour emergency information telephone number. Copies of Parts B and C were placed in the cockpit to be available to the flightcrew if required. Finally, copies of all the various hazardous materials paperwork were kept on file at the FedEx facility in Memphis.

With the preflight paperwork complete, the crew boarded their aircraft, N68055. The captain, forty-seven years old, was hired by FedEx in 1979, and had previously logged over 2,000 hours as a pilot in the DC-10. At age forty-one, the first officer had only 237 hours flying this type of aircraft. The flight engineer, forty-five, had been hired by the airline just six months earlier, accumulating only 188 total hours as a FedEx flight crewmember. Flight 1406 lifted off into the predawn darkness at 2:48 A.M. (CDT), six minutes late.

The flight proceeded uneventfully for almost two hours. At 5:36 A.M. (EDT), while level at 33,000 feet (FL 330), the cockpit voice recorder (CVR) captured the first indication of a problem. A conversation regarding a planned Category III landing procedure that was to be accomplished in Boston was interrupted by the illumination of a warning light.

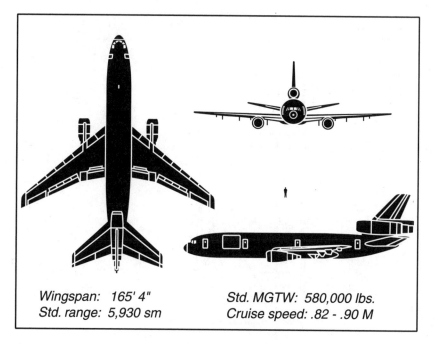

Wingspan: 165' 4"   Std. MGTW: 580,000 lbs.
Std. range: 5,930 sm   Cruise speed: .82 - .90 M

6-1 *Douglas DC-10-10.*

"What the hell's that?" asked the captain. Simultaneously the first officer and flight engineer responded "Cabin cargo smoke!"

Surprised by the warning during an otherwise normal flight, the captain stated again, "You see that? We got cabin cargo smoke...cabin cargo smoke!" But having been in an aircraft fire many years before, the captain immediately recognized the acrid smell and realized that "this was for real."

The flight engineer quickly started the emergency checklist, reciting "cabin cargo smoke, oxygen masks on...slash courier communications established." He then continued, "Okay, it's number nine smoke detector" (see Fig. 6-2 for smoke detector and cargo container locations).

All three flight deck crewmembers donned their oxygen masks and interphone communications were established. Neither the captain nor the first officer put their smoke goggles on, and the flight engineer removed his after only a short time because there was no smoke in the cockpit. The cockpit door was opened to visually check for smoke, but none was noted.

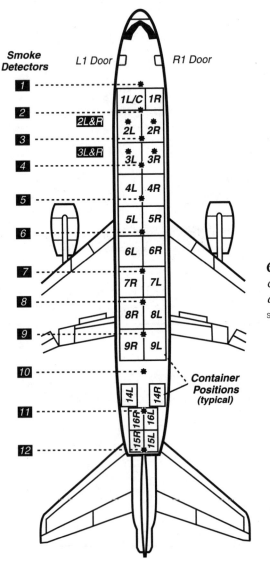

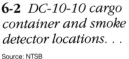

6-2 *DC-10-10 cargo container and smoke detector locations. . .*

Source: NTSB

Two company employees, one a B-727 flight crewmember and the other an international customer representative, were on board, riding as passengers in the "foyer" area immediately behind the cockpit (see Fig. 6-3). They were asked by the flightcrew to come forward. Anxious to get started with the emergency procedure, the captain then stated, "Okay, we're getting two of them [smoke detector illumination lights] now. Let's get on it..." Thirty seconds later his remarks became a little more ominous. "Okay, it's moving forward, whatever it is...it's up to [number] seven [smoke detector]."

For the next minute, the flight engineer continued with the checklist. "...Cockpit door and smoke screen closed...it's closed, if descent required proceed to step six...if descent not required proceed to step fourteen..." At one point, he stated "pull cabin air," referring to a control in the cockpit that regulates the flow of air into the cabin cargo area. Pulling the handle would restrict airflow, thus inhibiting the spread of the fire, but the "T-handle" was not pulled.

The captain asked for a test of the smoke detector system, but the results, blinking lights instead of steady ones, caused confusion among the crew. The flight engineer continued with the "cabin cargo smoke light illuminated" portion of the checklist, and at 5:40 A.M., only four minutes after the first indication of a problem, the captain stated, "We've definitely got smoke guys...we need to get down right now! Let's go!"

In the rush to get the aircraft in a descent and headed toward any nearby airport, the captain inadvertently keyed his air traffic control (ATC) radio microphone instead of using the flight deck interphone.

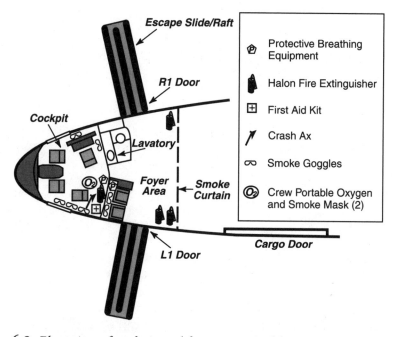

6-3 *Plan view of cockpit and foyer area, FedEx DC-10.* . . . Source: NTSB

This resulted in him broadcasting comments meant for other crewmembers on the ATC frequency.

"Okay, what's the closest field, I wonder...let's run it, let's get this thing depressurized, let's get it down."

The controller issued FedEx 1406 a turn to the left to fly directly to Stewart International Airport, Newburgh, N.Y. A maximum speed descent was initiated to 11,000 feet. "Okay, ready to run it [the emergency checklist] when you are," offered the flight engineer. "Okay," responded the captain. "Run the checklist."

"...Masks and goggles verify on one hundred percent...cockpit air outlets open...they are open...raise cabin altitude to twenty-five thousand..."

Twenty seconds later the flight engineer interrupted again with "and we now have detector eight, nine, and ten! We've lost detector seven, it's gone out." To obtain necessary landing data for Stewart Airport from the airplane's "airport performance laptop computer" (APLC), he needed to know its three-letter airport identifier. He asked, "Okay, what's that ah...stand by."

Meanwhile, the captain had confirmed with ATC his intentions of landing at Stewart, while flying directly there at maximum speed. The first officer asked the flight engineer, "...You have an approach plate for us?"

"What's the three-letter identifier for Stewart?" replied the engineer.

A brief discussion about the identifier ensued between ATC and the crew, and the engineer was distracted with other requests. "When you get a chance," asked ATC, "the...fuel on board and souls, please."

While the crew was receiving the current altimeter and other landing information from the controller, the engineer called out "it looks like we just have smoke detector ten lit now."

Twenty-eight miles from the airport, while the captain was clarifying with ATC the approach procedure to be used, the engineer again asked what the identifier for Stewart was. FedEx 1406 was cleared to descend to 4,000 feet and a only a few seconds later he stated again "three-letter identifier for that airport?" The captain replied "S-T-W," but since he was still transmitting over the radio, the controller corrected him with "Sierra whiskey foxtrot is Stewart."

"Okay, we are depressurized," stated the engineer, "...it says fire...check extinguished...the [smoke detector] lights are off...it's still smoky out there."

The captain confirmed that the airport's firefighting and rescue equipment would be standing by for their landing, and contacted the local approach controller. Headings were issued for a visual approach to Runway 27.

"I need the three-letter identifier for that airport so I can call it up..." Someone in the crew responded "S-W-F," and a few seconds later the first officer called for slats to be extended. They would begin their final approach in less than two minutes.

The emergency checklists were almost complete, but the engineer still had not given necessary landing data to the pilots. "What's the field elevation?" asked the first officer. The controller called at the same time offering approach course information, and the cockpit was simultaneously filled with the sound of the overspeed "clacker" and the altitude alert chime, alerting the crew that the aircraft was about to level at the assigned altitude. A descent to 3,000 feet was issued and acknowledged, and the intense workload elicited a "Boy, this sucks, doesn't it!" from the captain.

Just as the engineer asked one last time, "Is there a three-letter identifier—" the first officer requested any radio navigation information that would orient flight 1406 more precisely with the airport. A frequency and VOR radial were given, the altitude alert sounded once more in the descent, and the engineer finally stated, resignedly "I can't give you any takeoff or landing data."

"You can't?"

"I can't find the airport in my directory."

Then only five minutes from touchdown, the captain tried quickly to maintain control of the situation. "Just get a weight and use your table tops [landing speed information]...get rid of the boards [to configure the aircraft for landing]...three hundred thirty thousand pounds, V ref is one thirty-one for flaps fifty..."

The flight was cleared to a lower altitude and the "in-range" checklist initiated by the engineer. While the reference bugs on each airspeed indicator were being set at the proper speeds, the controller asked, "...the fire department needs to know if there are any hazardous materials on the plane." Getting an affirmative response from the engineer, the captain replied "Yes there is, sir."

Fourteen seconds later, a chilling remark was made. "And I've got additional smoke detectors on now!" The airport was then only ten miles away and the controller continued to issue landing advisories. The crew's workload was intense. The captain was still attempting to verify the final approach speeds with the engineer, the first officer called for additional flaps to help the aircraft to slow, and approach control issued yet another descent clearance.

As the landing gear was lowered and the checklist begun, the crew visually identified the runway only five miles from the threshold. Landing flaps were selected and the final checklist completed at 500 feet above the ground. At 5:54 A.M., eighteen minutes after the first smoke detector light illuminated, FedEx flight 1406 touched down at Stewart Airport.

"We need to get the hell out of here!" yelled the captain after the aircraft came to a stop. Just then an engine fire alarm blared in the cockpit. "Blow, blow the door!" commanded the captain. That was the last sound recorded on the CVR.

A dense gray haze filled the foyer area. Opening the cockpit door, the flight engineer could not even see the smoke barrier (see Fig. 6-3), a curtain separating the foyer area from the cabin cargo compartment. All three engine fire control handles were pulled, shutting off fuel and hydraulic fluid to the engines and disconnecting electrical power. All of the internal engine fire extinguishers were discharged.

After exiting the flight deck, the flight engineer attempted to open the two forward main doors, L1 and R1. Neither would budge. In unsuccessfully trying to open his sliding cockpit window, the captain realized that the aircraft was still pressurized. All the windows and doors on the DC-10 were the "plug" type, requiring inward initial movement, and any internal pressure greater than that outside the airplane would prevent their opening. In August of 1980, 301 lives were lost after another three-engine wide-body aircraft, an L-1011, landed with an onboard fire and internal cabin pressure prevented any exits from being opened.

After breaking the air seal on the window, smoke rushed into the cockpit. "We're still pressurized! Get back on oxygen!" shouted the captain.

After manually rotating the pressurization "outflow valve" controller to the full open position and relieving all internal pressure, the engineer was able to open the R1 door. The slide deployed automatically, and the engineer and both couriers evacuated the airplane. Both cockpit windows were opened, and with acrid smelling smoke billowing out of the aircraft, both the captain and first officer maneuvered out of the windows and down escape ropes to the ground.

Emergency response

As promised, six ARFF (aircraft rescue and fire fighting) Air National Guard fire trucks moved into position around flight 1406 as the aircraft came to a stop on taxiway A3 adjacent to the runway (see Fig. 6-4). The flight engineer gave the first fireman he saw the cover sheet of Part A of the "Notification of Dangerous Goods," which he had retrieved from the cockpit before evacuating. He also informed him of the position of known dangerous materials on board the aircraft and the fact that "Part B" of the form, containing additional detailed information, was located

6-4 *FedEx 1406 burning on the taxiway shortly after landing.* . . Courtesy NTSB

in a plastic folder on the cockpit door. That information was then radioed to the incident commander, the assistant fire chief on duty.

Initial attempts to control the blaze focused on the use of handheld hose lines by firefighters who had used a ladder at the L1 door to get into the aircraft. Congregating in the foyer area, they found that the cargo net and forward cargo containers blocked access to the fire, located much further aft in the airplane. Difficulty was also encountered opening the main cargo door. The FedEx station manager, a company mechanic, the flight engineer, and others finally raised it about fifty minutes after landing, but five minutes later flames had burned through the top of the fuselage.

The use of firefighting equipment with a penetrating nozzle (a "skin penetrator agent application tool," or SPAAT) by the fire department was considered but not initially used due to possible damage it would cause to the aircraft. By the time the time the decision was made and the tool was rigged and ready for use, the fire had compromised the structure of the aircraft. At that point, the incident commander recognized the danger posed to his men, and removed all firefighters from the interior of the aircraft. Truck-mounted turrets aimed fire extinguishing foam at those areas of flame coming through the fuselage, but in spite of all their efforts and over 50,000 gallons of water and firefighting foam, the airplane continued to burn until about 9:25 in the morning, three-and-a-half hours after landing.

The aircraft and most of the cargo had been destroyed, with total losses placed at $300 million (U.S.). The fuselage had separated just behind the trailing edge of the wing and also at the aft pressure bulkhead (see Fig. 6-5).

The investigation and findings

National Transportation Safety Board (NTSB) investigators from the Northeast Regional Office augmented a go-team from Washington, D.C. Specialists in fire, hazardous materials, operations, airworthiness, and air traffic control were included, with the primary goal of the team to determine the origin of the in-flight fire.

Origin of fire

An extensive survey was done of the cabin cargo area and all cargo containers in an attempt to locate the origin of the fire. This task was made much more difficult by the fact that the fire had burned for

6-5 *FedEx 1406—the aftermath. . .* Courtsey Joe Bracken, ALPA

about four hours under changing conditions, destroying potentially important evidence.

The most severe heat and fire damage was found in and around container 6R. Every other container except that one exhibited a layer of unburned cargo that covered the integral floor. Additionally, this container was the only one with heat damage on the bottom surface and extensive scorching of the composite flooring material. The Board found that "the deepest burned-out area [was] centered over container 6R and that the cargo containers surrounding 6R and all the contents in these containers were all burned to a greater depth along the sides common to container 6R."

There was also heat damage in the area just aft of container 9L, but no cargo was loaded in that position on flight 1406. Investigators determined that burning material from the fuselage crown or possibly from the container in 9L probably caused that damage. Furthermore, container 9L had a "significant" amount of unburned materials, whereas in 6R the "vast majority of easily burned unprotected material...was consumed by the fire."

One piece of evidence of interest to the Board was a "V" burn pattern that originated at 6R. Often, this type of pattern will point to the origin of a fire, but not always. According to the National Fire

Protection Association's (NFPA) publication 921 *Guide for Fire and Explosion Investigations,* ventilation fluctuations and changes in fuel sources can produce additional "V" patterns. "Determining which pattern was produced at the point of origin by the first material ignited becomes more and more difficult as the size and duration of the fire increases."

There were varying witness observations regarding the location of the first flames to penetrate the fuselage. Some indicated it was five to six feet behind the leading edge of the left wing, corresponding to an area near the front of the 6L container. Others indicated a location above the area occupied by containers 8 and 9. Comments recorded on the CVR might also indicate an origination in that area, as the earliest smoke detector activated was number 9. But the NTSB pointed out that the "breakthrough" principle, the premise that the initial fire source is near the area of flames that first break through the aircraft structure, was not always valid. Three accidents were pointed out as examples of this anomaly. A ValuJet DC-9 fire at Atlanta in 1995, an Air Canada DC-9 in-flight fire and subsequent landing at Greater Cincinnati Airport in 1983, and a Delta 727 ground fire at Salt Lake City in 1989 all demonstrated fuselage flame breaches in areas well away from the fire source.

Other information provided to the Board by various fire experts indicated that the smoke detector system on the DC-10 is only to detect the presence of fire or smoke, and was never designed to pinpoint fire locations. The Board summed up its difficulty of tracking down the location of the fire's origin, stating: "...there was insufficient reliable evidence to reach a conclusion as to where the fire originated."

Ignition source of the fire

Most investigators believed the "V" burn pattern and the severity of heat damage found at the 6R location to be very significant, thus requiring further scrutiny. In the container itself they found a pallet of industrial metal valves, a Texas Instruments laptop computer, two Power Station power servers, a Power Computing computer, miscellaneous computer parts, and an Expedite Model 8909 DNA synthesizer manufactured by PerSeptive Biosystems. All were examined carefully.

The computers and power servers exhibited external melting and burning, but little or no internal fire damage. Internal batteries were either undamaged or discharged. The shipment of metal valves was charred and the pallet was singed and burned.

The DNA synthesizer, owned by Chiron Corporation in California and being shipped to a subsidiary in Massachusetts, was found in the center of 6R, lying on its side and at the lowest point of the "V" burn pattern. The unit (see Fig. 6-6) was designed to produce synthetic DNA from a variety of chemical reagents and was housed in a metal cabinet about thirty inches high. The primary reagents used in the process were acetonitrile and tetrahydrofuran (THF), both of which are classified as flammable liquids by Department of Transportation (DOT) regulations. Acetonitrile has a flash point of forty-two degrees (F) and THF can form peroxides that explode on contact with strong bases or metals, or even spontaneously at higher concentrations.

There was an access door with a broken glass viewing panel on the front of the cabinet, and fifteen brown reagent bottles of varying sizes were found inside, still screwed into caps attached to the internal mechanism of the synthesizer. A few still had liquid in them, several had flammability symbols printed on their labels, and one had a particularly strong chemical odor. Four large external reagent bottles necessary for proper operation of the synthesizer were not included in the shipment. After carefully studying the evidence, the Board pointed out, "The nature and degree of the fire damage to the synthesizer...was thought to be suggestive of a source of fuel inside the synthesizer."

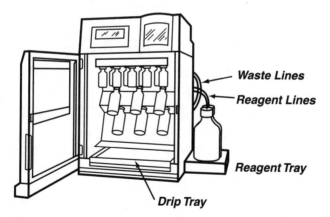

6-6 *Expedite model 8909 DNA synthesizer instrument cabinet. . .* Source: NTSB

In December of 1996, all of the fluids found in the machine were analyzed at the NASA Kennedy Space Center. Every bottle still contained volatile, flammable chemical reagent chemicals. Traces of firefighting agent also indicated that the bottles had been open to the atmosphere at some point.

Regulations dictated that prior to shipment of a DNA synthesizer, an extensive purging procedure must be completed to completely remove all flammable chemicals from the machine. For comparative purposes, a demonstration was conducted at the Armed Forces Institute of Pathology (AFIP) in February of 1997, using an identical synthesizer and following PerSeptive's approved purging procedures. When the purging was complete, all reagents remaining in the machine were measured and analyzed. The Safety Board discovered that bottles from the accident synthesizer contained from two-and-a-half to five times more acetonitrile and THF than did the bottles from the properly purged demonstration machine. The Board also pointed out that "it is likely that significant amounts of the chemicals [acetonitrile and THF] were consumed in the prolonged and intense fire and thus the synthesizer probably contained much larger quantities of these flammable chemicals before the fire."

Evidence suggested that the PerSeptive field engineer responsible for preparing the accident synthesizer at the Chiron laboratory experienced difficulties in properly purging the unit. The Safety Board noted, "The most reasonable explanation for the presence of excessive quantities of chemicals in the synthesizer is that one or more of the bottles...was not sufficiently emptied before the purging process began." They concluded that "the DNA synthesizer was not completely purged of volatile chemicals (including acetonitrile and THF) before it was transported on board flight 1406." Investigators were unable, however, to develop to everyone's satisfaction a complete scenario to explain how the synthesizer started the fire.

Further examination of other cargo shipments on board, particularly in the area of container rows 8 and 9, yielded no significant findings. Four separate shipments of marijuana were discovered in the wreckage, totaling almost one hundred pounds, but were ruled out as a possible ignition source because all were vacuum packed and exhibited no signs of spontaneous combustion. Attempts to trace those packages back to an individual or shipper were unsuccessful. Finally, there was no evidence to suggest that any of the airplane's systems experienced a malfunction that would have led to a fire.

Once again, findings were inconclusive. "In sum, the Safety Board could not conclusively identify an ignition source for the fire."

Flightcrew action

Although a successful emergency landing was accomplished, the NTSB identified several areas of questionable flightcrew performance. Initially, several items on the "Cabin Cargo Smoke Light Illuminated" checklist were not completed. These included the failure to pull the T-handle to shut off air to the main deck cargo area, thus allowing the fire to intensify, and adequate depressurization of the aircraft, preventing a rapid emergency evacuation. During the in-flight portion of the emergency, the flight engineer was overloaded with tasks and distracted by his inability to obtain an identifier for the airport. As the Board noted, the captain allowed the first officer to continue flying the airplane, thus attempting to apportion the workload among the crew, but never called for any checklists, nor did he assign specific duties to other crewmembers.

The Board found that the captain "failed to provide sufficient oversight and assistance to ensure completion of all necessary tasks." Furthermore, they concluded that he "did not adequately manage his crew resources when he failed to call for checklists or to monitor and facilitate the accomplishment of required checklist items."

Dissemination of hazardous materials information

As with any firefighting effort involving hazardous materials, timely information regarding the quantity and location of those materials on board an aircraft is critical in order to adequately fight the fire and protect emergency response personnel. In this case, the incident commander at the Stewart Airport received only the Part A form and a handwritten list provided by FedEx employees on the field. Critical information was severely limited, with partial identification of some chemicals on board received only ten minutes before the fire was extinguished.

After the accident, FedEx had faxed many shipping documents to the Emergency Operations Center (EOC) at the airport, but most were illegible and none were ever received by the incident commander. Additionally, the Safety Board found that "FedEx did not have the capability to generate, in a timely manner, a single list indicating the shipping name, identification number, hazard class, quantity, number of packages, and the location of each declared shipment of hazardous materials on the airplane." There were eighty-five separate hazardous materials packages on board, and the procedure used of faxing copies of each of the Part Bs proved to be "burdensome, time consuming and, in this case, ineffective. Also, because of the poor quality and legibility of many of the handwritten Part Bs, much of the information was unusable," according to the Board.

The organizations initially involved in responding to the accident were the New York State Police, the Orange County Office of Emergency Management, the county Hazardous Materials Response Team, the Department of Environmental Conservation, Stewart International Airport, the Air National Guard, and FedEx. Each had previously conducted exercises to prepare for hazardous materials incidents, but none were done in conjunction with the other organizations. The NTSB found that "...communication and coordination among the participating agencies were not effective." Further, inadequate planning resulted in "confusion about the responsibilities of [each] agency and contributed to the failure of information about the hazardous materials on the airplane to reach the incident commander."

Conclusions and probable cause

The NTSB found that the flightcrew and aircraft were properly certificated, ATC assistance was appropriate and timely, and the smoke detector system on board the aircraft functioned as intended. Weather was not a factor.

Materials discovered in the cargo of the accident flight, such as aerosol cans, acidic liquids, and marijuana, indicated that common carriers can be unaware of the true contents of the packages they carry. As NTSB Chairman Jim Hall pointed out during the hearing, "hidden shipments (of undeclared hazardous materials) still pose a major problem" for the aviation industry. "There's a lot of hazmat criss-crossing our skies every day," he said. Also, DOT regulations do not address problems discovered with dangerous goods information retrieval and better hazardous material emergency handling plans are needed. Finally, the Board noted that inadequate means exist for combating on-board aircraft fires.

The NTSB determined that "the probable cause of this accident was an in-flight fire of undetermined origin."

Recommendations

The investigation produced the following recommendations:

To the Department of Transportation:

- Require, within two years, that written responses to inquiries regarding hazardous materials characteristics of cargo be included in the shipping papers and that methods be explored to improve detection of undeclared hazardous materials. (A-98-71)

To the FAA:

- Require the principal operations inspector (POI) of FedEx to review the crew's actions to determine if changes to company procedures and training are necessary. Require the POI to ensure that all FedEx employees are aware that hazardous material information must be provided to personnel responding to an accident. (A-98-72 and 76)

- Require FedEx to modify its checklists to emphasize the availability of protective breathing equipment for crewmembers during emergency evacuations. The Board also re-emphasized previous recommendations calling for guidance to air carrier pilots about the urgent need to don oxygen masks and goggles early, and the establishment of performance standards for rapid donning of smoke goggles (A-98-73, A-97-58 and 59). These earlier recommendations stemmed from the ValuJet 592 accident in the Florida Everglades in May, 1996 (see Chapter 5, "Carnage in the Everglades").

- Require all operators of aircraft with doors similar to the DC-10 to remind crewmembers of the need for aircraft depressurization before attempting to open the door. (A-98-74)

- Require all airports to develop a coordinated hazardous materials response plan that specifies the responsibilities of each agency and to schedule joint exercises to test these plans. (A-98-77)

- Reexamine the feasibility of on-board aircraft cabin interior fire extinguishing systems and review firefighting policies and procedures to develop improvements in training and equipment. (A-98-78 and 79)

To the FAA and Research and Special Programs Administration (RSPA):

- Require, within two years, that air carriers have the means to quickly retrieve and provide to emergency responders specific hazardous material information, including proper shipping name, hazard class, quantity, and location on the aircraft. (A-98-75 and 80)

Industry action

Shortly after the accident, FedEx acted to revise and clarify the "Fire and Smoke" and "Cabin Cargo Smoke Light Illuminated" checklists. A new introductory paragraph explains the reasons for the procedures, several steps were emphasized, and new ones were added to assure aircraft depressurization and retrieval of appropriate cargo documentation from the cockpit. A new "Quick Evacuation" checklist was developed, incorporating only the most important, immediate action items.

Flightcrew training was modified to include the addition of a new in-flight fire scenario in the normal training syllabus. Procedures to eliminate smoke and fumes from the cockpit and proper crew resource management techniques are emphasized. New full-face pilot smoke masks have been purchased for the entire fleet, providing better user protection and enhanced visibility.

FedEx also initiated development of a comprehensive system of hazardous material tracking and information retrieval that should be in place by the year 2000. Initially, electronic notification of hazard class, quantity, and location of hazardous cargo would be available on any FedEx computer terminal. Eventually, all this information and more, including complete hazardous material manifests, will be available at the FedEx Global Operations Command Center (GOCC) at all times. The company believes, however, that current regulations will need to be adjusted to allow electronic record keeping in lieu of the paper documentation now required.

The FAA is currently in the process of writing a flight standards information bulletin directing POIs to remind flightcrews of the importance of complete aircraft depressurization before attempting to open doors. Guidance contained in Advisory Circular (AC) 150/5200-31, *Airport Emergency Plan*, used by airport management personnel, is being updated, particularly in the handling of hazardous materials incidents.

While the FAA continues to explore the feasibility of other on-board cabin fire protection equipment such as water spray systems, there are no plans to initiate any applicable rule-making activity. The FAA has developed new firefighting equipment, specifically a truck-mounted elevated boom with a fuselage skin penetrating nozzle, now used at many airports in the U.S. They have also funded twelve large firefighting-training academies throughout the country, each with an aircraft interior fire simulator. Finally, the RSPA is drafting a Notice of Proposed Rulemaking (NPRM) requiring air carriers to develop and implement their own comprehensive system of hazardous materials information retrieval and distribution.

Epilogue

The Safety Board's investigation into this accident was comprehensive, uncovering shortcomings in the system that led to appropriate safety recommendations and the enhancement of aviation safety. But it was a difficult process.

Those corporations that feared liability refused the normal process of informal witness interviews, requiring that depositions be taken under strict guidelines. While beneficial to the investigation, information gleaned in an antagonistic setting is not always as straightforward as it might otherwise be.

Regulatory guidelines allow the NTSB to include qualified, interested "parties" in their investigation because of the inherent expertise they can make available to the Board (see Introduction). In this investigation, a manufacturer and an owner of equipment being shipped on flight 1406 hired experts outside the party's area of expertise whose testimony was included in the Board's reports. This practice was a deviation from accepted NTSB policies, and may not have been in the best interests of the investigation.

The Independent Safety Board Act of 1974 established complete autonomy for the NTSB to give them the structure and protection needed to find the "causes" of accidents, which they do well. But it should be up to those outside the investigation to worry about "blame."

References and additional reading

Aarons, Richard N. 1998. Cause and Circumstance: Fire Drill. *Business & Commercial Aviation*. March, 140.

Evans, David, Managing Editor. 1998. Systemic safety shortcomings found in investigation of cargo jet fire. *Air Safety Week*. 12(30).

Federal Aviation Administration. 1998. *NTSB Recommendations to FAA and FAA Responses Report*. <http://nasdac.faa.gov>

Flight Safety Foundation. 1998. After smoke detected in cargo compartment crew lands DC-10, then fire destroys aircraft. *Accident Prevention*. November, December, 20.

National Transportation Safety Board. 1997. *Aircraft Accident Report: In-flight Fire/Emergency Landing, Federal Express Flight 1406, Douglas DC-10-10, N68055, Newburgh, New York, September 5, 1996. NTSB/AAR-98/0*. Washington, D.C.: NTSB.

National Transportation Safety Board. 1996. Public Docket. *Federal Express Flight 1406, Douglas DC-10-10, N68055, Newburgh, New York, September 5, 1996. DCA96MA079*. Washington, D.C.: NTSB.

Phillips, Edward H. 1996. Safety Board intensifies probe of FedEx DC-10 fire. *Aviation Week & Space Technology*. 16 September, 92.

7

Contact approach

A close call on Delta flight 554

Operator: Delta Air Lines

Aircraft Type: McDonnell Douglas MD-88

Location: LaGuardia Airport, New York

Date: October 19, 1996

By all accounts, the weather in the northeastern United States was miserable. Portions of Philadelphia and New York were flooded from the effects of heavy rains and high winds, and New England was receiving a pounding as an intense low-pressure system moved to the northeast. Aircraft approaching these destinations were encountering low ceilings and visibilities, along with bumpy flight conditions.

About 4:15 in the afternoon, in anticipation of their planned landing at New York's LaGuardia Airport, the first officer of Delta flight 554 made a farewell public address (PA) announcement to his passengers. "Ladies and gentlemen, one more update from the flight deck. Right now we are about fifteen miles from the airport at LaGuardia and we expect to start our approach pretty shortly...The latest weather reports are, it's still raining at the airport and the winds are out of the southeast. The velocity has decreased somewhat. It's now fifteen to twenty miles an hour instead of the thirty to thirty-five they had before. As we go through the clouds and the rain we will get a few bumps as we make our approach, and it's probably gonna be a little bit bumpy all the way until landing..."

And with that statement, probably no one on that aircraft envisioned just how bumpy the landing would be. Or just how close their flight would come to ending in a total disaster.

Flight history and background

About 11:45 on the morning of the accident, the forty-eight year old captain was notified by Delta Air Lines crew scheduling of an assignment to fly a three-day trip. The first leg would be Delta flight 554, nonstop McDonnell Douglas MD-88 (see Fig. 2-1) service from Atlanta Hartsfield International Airport (ATL) to New York's La-Guardia (LGA). He and the first officer reported for duty at around 1:30 P.M. (EDT). As part of their preflight duties, they reviewed a company-provided weather package, which contained advisories for moderate-to-severe turbulence in strong winds at their destination.

The flight departed the gate at 2:31 P.M. and took off about ten minutes later with the captain at the controls. He had joined Delta eighteen years earlier and had 10,000 flight hours, 3,700 of which were as pilot-in-command of the MD-88. He completed a FAA first class medical examination just eleven days before the flight, and his medical certificate contained the restriction, "Must have glasses available for near vision." The first officer was ten years junior to the captain, both in age and in airline seniority. He had about 6,800 flight hours, including 2,200 hours as second-in-command in the MD-88.

The departure, climb, and en route portions of the flight were routine. However, as Delta 554 approached the New York terminal area the crew observed large areas of precipitation on the aircraft's weather radar, and the flight encountered light-to-moderate turbulence. Having been briefed to expect a bumpy ride, the flight attendants had prepared the cabin for landing and were seated in their jump seats earlier than usual. As one of the flight attendants made a PA announcement to ask passengers to prepare for landing, the captain remarked to the first officer, "Hang on to your hats folks, ladies and anybody else."

Before contacting New York approach at 4:12 P.M., the crew had obtained Automatic Terminal Information Service (ATIS) "Delta" which indicated that the airport visibility was one and one-quarter miles in heavy rain and mist, with an overcast ceiling at 1,300 feet above ground level (agl). When the captain heard the current altimeter setting of 29.50 inches of mercury, he commented to the first officer, "That's a pretty stout low pressure!"

During the crew's approach briefing for the Instrument Landing System (ILS) approach to LaGuardia's Runway 31 the captain remarked that the localizer beam was offset three degrees from the runway centerline. He also read a note on the approach chart that stated that the glide slope was unusable below 200 feet mean sea level (msl) (as depicted in Fig. 7-1).

At 4:33 P.M., New York Approach cleared Delta 554 for the ILS Runway 13. At about the same time, the captain remarked to the first officer, "It's gonna look real funny, plus it's gonna be right behind this post. [That] is where the runway's gonna be." Because of the wind correction angle that they were holding during the approach, the captain was highlighting that the runway would be angled off the nose, and that it would be somewhat obstructed behind one of the windshield posts. He then stated, "I may need the windshield wipers on final."

Shortly thereafter, the crew contacted LaGuardia Tower. "Delta five five four, you're number two," replied the tower. "Traffic to follow, two mile final. The wind now one zero zero at one two. One departure prior to your arrival. Braking action good by a seven thirty seven. Low-level wind shear reported on final by a seven thirty seven." The Delta crew then completed their landing checklist.

When the aircraft was approximately 1,500 feet above the ground, LaGuardia Tower radioed that the Runway 13 touchdown runway visual range (RVR) was 3,000 feet and the rollout RVR was 2,200 feet. "Must be raining hard at the airport," the captain remarked.

"Alright, a thousand feet above minimums," called the first officer. The captain responded that he did not see the runway yet, and asked, "What was the ceiling on the ATIS?"

"Ah, thirteen hundred," replied the first officer.

"You can forget that," retorted the captain.

At 4:37:13 P.M., LaGuardia Tower cleared a TWA jet for takeoff on Runway 13, and then cleared Delta 554 to land. The first officer told the captain, "Starting to pick up some ground contact."

Within seconds, though, the TWA crew notified LaGuardia Tower that they needed to reject the takeoff and would be turning off the runway. Almost immediately, the Delta captain became concerned that TWA's rejected takeoff would mean that they would have to

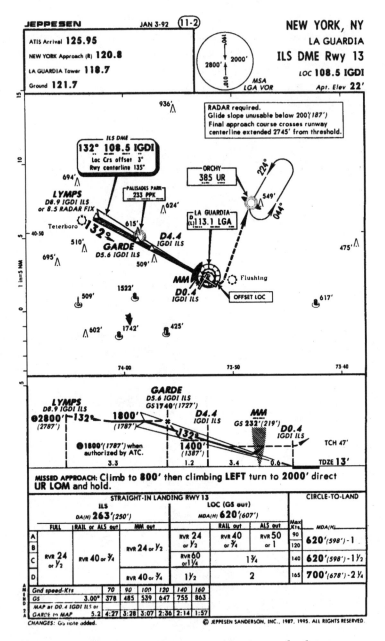

7-1 *LaGuardia ILS DME Runway 13 approach plate as reproduced in the public docket...* Source: NTSB

abandon their approach due to the possibility of TWA still being on the runway. He disconnected the MD-88's autopilot and stated, "I've got the jet."

Delta 554 continued down the approach path as TWA taxied clear of the runway. At 4:37:57 P.M., the Delta first officer stated, "Two hundred above [minimums]," followed by "speed's good, sink's good."

"No contact yet," the captain said.

"One hundred above," called the first officer.

"I got the REIL. Approach lights in sight," stated the captain, indicting his visual contact with the Runway End Identification Lights (REIL) and approach lights.

"You're getting a little bit high," noted the first officer. "A little bit above glide slope. Approach lights. We're left of course."

As the aircraft descended to the Decision Altitude (DA) an onboard alerting system verbally announced "Minimums." The MD-88's windshield wipers were turned to full speed and the captain reiterated, "Approach lights in sight."

In response to being high, the captain disconnected the autothrottles and manually reduced thrust on both engines. "Speed's good. Sink's seven hundred," called the first officer. The captain then said, "I'll get over there," indicating that he was turning to more precisely align with the runway.

Almost immediately, at 4:38:31.1 P.M., the first officer cautioned, "A little bit slow. A little slow." The captain began adding thrust to the two Pratt and Whitney engines and simultaneously pitching the aircraft nose up. Two seconds later, the first officer exclaimed, "nose up!" The GPWS called, "sink rate," and the first officer again immediately exclaimed, "nose up!" At 4:38:36.5 P.M., the GPWS again announced, "sink rate." Eight tenths of a second later the Cockpit Voice Recorder (CVR) recorded the sound of impact.

Runway 13 is constructed on an elevated deck about twenty feet above the Flushing Bay. The extended portion of the runway is constructed of asphalt and concrete that is laid out on steel piers, with its approach end covered by orange and white plywood panels that extend vertically towards the water. Traveling at around 130 knots, the MD-88 first hit two approach lighting stanchions and the over-water catwalk that connected them. Both main landing gear then impacted the vertical edge of the concrete runway deck, where both main landing gear trucks were sheared off and fell into the bay. The crippled aircraft then skidded on its lower fuselage and nose landing gear approximately 2,700 feet down Runway 13 before finally coming to a stop. The nose wheel came to rest on the runway pavement, but during the ground slide the fuselage had turned completely around, facing the approach end of the runway with the right wing hanging over a wet grassy area next to the runway.

After the aircraft stopped, the flightcrew began to assess aircraft damage and determine whether an emergency evacuation was necessary. About seventy-four seconds after the aircraft came to rest, the captain ordered an evacuation, when a nonrevenue Delta pilot who had been sitting in the passenger cabin reported smelling jet fuel fumes. All fifty-eight passengers and five crewmembers exited through the front left door using the escape slide. Three passengers reported minor injuries.

The investigation and findings

The NTSB responded with personnel from its Northeastern Regional Office, supplemented by expertise from their Washington, D.C., headquarters. Because the accident forced closure of LaGuardia Airport, those traveling to the field investigation were forced to rely on surface transportation. One party coordinator found himself on a five-hour rainy, bumpy bus trip from Boston to New York.

Investigators found damage to the approach lights in two areas several hundred feet from the approach end of the runway. The plywood covering the vertical portions of the runway deck exhibited two main areas of impact damage, corresponding to the exact dimensions of the main landing gear and wheels (see Fig. 7-2). Scraping and scaring of the runway was minimal because the runway was very wet with areas of standing water, although pieces of aircraft and approach light structures were scattered along the wreckage path.

The aircraft right wing exhibited extensive denting, crushing, and tearing, with wood and fiberglass embedded in the leading edge. The right wing fuel tank was punctured, allowing an estimated 600 gallons of fuel leakage on the ground beside the runway. The underside of the fuselage structure was extensively damaged, with

7-2 *Approach end of LGA Runway 13. Note impressions of both main landing gear trucks in plywood panels and undamaged obstruction light between...*Courtesy ALPA

crushed stringers, frames, and longerons. Examination of the remains of the landing gear yielded a chilling discovery: the gear struts were sheared-off about twelve inches below the bottom surface of the wing. Had the aircraft been a foot lower when it contacted the vertical portion of the runway deck, no doubt the outcome of this accident would have been much more severe.

Weather

Examination of the LaGuardia Tower ATC transcripts revealed that about an hour before the accident, between 3:25 and 3:45 P.M., there were numerous pilot reports of wind shear. Further, during this period four flights executed missed approaches due to wind conditions while attempting to land on Runway 13. However, between 3:45 P.M. and the time of the accident there were no pilot comments regarding wind shear.

Automated weather observations taken at LaGuardia Airport around the time of the accident indicated that the surface winds were generally out of the east, with velocities fluctuating between eleven and sixteen knots. It was raining heavily and surface visibility was between one-half and three-quarters of a mile. Weather data were also obtained from a local National Weather Service (NWS) Doppler Radar. Using this information a vertical wind profile was created that showed winds between 1,000 and 5,000 feet agl were constant in direction (winds out of the east) and velocity (sixty to seventy knots.)

Examination of Delta 554's flight data recorder (FDR) showed no sudden airspeed or altitude changes associated with wind shear. The CVR did not contain any information regarding the flight encountering wind shear, either through crew comments or through activation of the onboard wind shear computer that would have provided aural alerting to the crew. Post-accident examination of the onboard wind shear computer indicated the system was operational and capable of generating and annunciating these alerts.

The NTSB also inspected the FDRs of three of the four flights that preceded Delta 554 to Runway 13. These flights, a Delta 727, a Continental 727 and a USAir 737 landed at 4:30, 4:33, and 4:34 P.M., respectively. The Safety Board also obtained the FDR of the United 737 that was immediately behind Delta 554 on the approach when the accident occurred, and was directed to go-around by the tower while they were still at 1,800 feet agl. Examination of the FDR data and pilot reports from these four aircraft revealed no evidence of wind shear encounters during their approaches to Runway 13. Using all of the above weather information, the Safety Board concluded that Delta 554 did not encounter wind shear during its approach.

Airport issues

The elevation of LaGuardia Airport is twenty-two feet msl. Operational considerations and prevailing winds usually preclude air traffic control from utilizing Runway 13 for landing. Because the first several hundred feet of the runway are on an overwater pier, glide slope signal irregularities occur due to tidal changes and the metal content of the water. Therefore the approach could only be certified with the restriction that the glide slope be noted as unusable below two hundred feet msl. The day after the accident the FAA conducted an airborne operational check of the ILS, and all components were determined to be operating normally.

Runway 13 was also served by a Visual Approach Slope Indicator (VASI) light system, providing visual descent guidance information to pilots during an approach to the runway. According to the pilots of

Delta 554, when they descended below two hundred feet agl and glide slope guidance was considered unusable, they were in visual flight conditions. However, despite being in visual conditions they did not observe any of the VASI lights during the descent to the runway. Additionally, post-accident interviews with pilots of the four aircraft that landed immediately before flight 554 could not recall observing the VASI lights during their final approach and landing. However, the Safety Board noted that "none of the pilots interviewed (including the flightcrew of flight 554) recalled specifically seeking VASI light guidance during their approach to land." In fact, the NTSB's review of LaGuardia Airport facility records and interviews with ATC personnel could uncover no indications that the VASI lights were impaired or otherwise inoperative during the period that flight 554 was attempting to land.

Investigators learned that the runway edge lights alongside Runway 13 were installed with irregular distances between them, ranging anywhere from 120 to 170 feet between lights. The NTSB noted that the FAA's criteria for runway edge light spacing specifies that these lights must be no further apart than two hundred feet, and that they should be spaced "as uniformly as possible." Observing that most airports have runway edge lights spaced at or near the FAA-established maximum distance of two hundred feet, the NTSB found that the spacing on these lights at LGA varied even when no physical limitations existed, such as intersecting runways or taxiways.

The airport was found to be fully compliant with 14CFR Part 139, "Certification and operations: Land airports serving certain air carriers." Because the approach end of Runway 13 is built on a pier, it does not have the normally required runway safety area as specified in Part 139. It is not required to, however, because all runways built prior to 1988 have been "grandfathered" from complying with the regulation. There is no doubt, though, that had a safety area existed for Runway 13, there would have probably been very little, if any, damage to the aircraft.

The aircraft

The accident aircraft, N914DL, was placed into service by Delta in June 1988. It was powered by two Pratt & Whitney JT8D-219 turbofan engines. No noteworthy discrepancies were found in the maintenance log, and the pilots did not indicate any maintenance irregularities during their flight from Atlanta.

The MD-88's vertical speed indicator

The accident aircraft was equipped with a "noninstantaneous" vertical speed indicator (VSI). With a noninstantaneous VSI, if the aircraft's vertical speed (rate of climb or descent) changes rapidly, it can take up to four seconds before the instrument indicates the actual climb or descent rate. The aircraft manufacturer stated that the VSI could be updated to display instantaneous vertical speed information if an inertial reference unit (IRU) was installed in the airplane in place of the attitude/heading reference system (AHARS). At the time of the accident, Delta was replacing AHARS with IRUs, although the accident aircraft had not undergone this conversion. The NTSB interviewed several Delta MD-88 check airmen and flight instructors who believed that most Delta line pilots were unaware that the VSIs in the MD-88s were not instantaneous.

The NTSB determined that the VSI's lag time "limited the first officer's ability to provide the captain with precise vertical speed information during the critical final seconds of the approach, and therefore contributed to the accident."

Flight data recorder information

The FDR was read out and analyzed at the NTSB's Washington, D.C. laboratory. Figure 7-3 shows a synopsis of FDR data from the last sixty-three seconds of flight, with CVR excerpts superimposed with FDR information.

The aircraft was established on the electronic glide slope and localizer until it reached about 400 feet msl, where it began to deviate above the electronic glide slope. At 4:38:20 P.M., the point where the first officer called "minimums," the aircraft was 1.43 dots above the glide slope, and five seconds later it was 2.40 dots high. The first officer stated "speed's good," as the captain had previously briefed that 131 knots would be the target approach speed. At that point, the engine's thrust, or engine pressure ratio (EPR), was being reduced from about 1.2 EPR to about 1.15 EPR. That coincided with the point where the aircraft was descending below 200 feet msl and the electronic glide slope would no longer be usable.

At 4:38:26 P.M., the first officer called "sink's seven hundred," indicating a 700 foot-per-minute (fpm) descent rate. However, realizing the inherent lag with noninstantaneous VSIs, the NTSB used FDR information to calculate that the actual rate of descent was

CVR Excerpt	Local time	msl alt	Radio alt	IAS knots	G/S dev. dots	LOC dev. dots	EPR eng. 1	EPR eng. 2	Elevator pos. trailing edge up
A/P off	4:37.33	730	895	131	0.09 fly down	0.03 fly right	1.26	1.26	6.4 degrees
200 above	4:37.57	468	603	129	0.06 fly up	0.04 fly right	1.27	1.27	7.5 degrees
100 above	4:38.10	377	465	131	1.18 fly down	0.26 fly left	1.30	1.28	7.0 degrees
approach lights in sight	4:38.11	376	453	131	1.30 fly down	0.32 fly left	1.33	1.32	5.2 degrees
little bit high	4:38.13	341	435	130	1.39 fly down	0.39 fly left	1.37	1.35	7.0 degrees
minimums	4:38.20	265	319	133	1.43 fly down	0.67 fly left	1.19	1.21	9.3 degrees
speed's good	4:38.25	213	259	129	2.40 fly down	0.87 fly left	1.16	1.19	4.8 degrees
sink's 700	4:38.26	202	239/189	126	2.32 fly down	0.91 fly left	1.13	1.15	9.5 degrees
I'll get over there	4:38.30	151	143	127	0.89 fly down	0.84 fly left	1.09	1.09	4.3 degrees
a little bit slow...	4:38.31	133	118	126	0.27 fly down	0.79 fly left	1.08	1.08	8.8 degrees
nose up	4:38.33	68	59	124	2.08 fly up	0.71 fly left	1.23	1.11	20.3 degrees
nose up / sink rate	4:38.34	39	30	124	3.16 fly up	0.65 fly left	1.48	1.18	24.9 degrees
GPWS "sink rate"	4:38.35	23	10	126	4.03 fly up	0.55 fly left	1.65	1.43	26.0 degrees

7-3 *Selected comments from CVR correlated with FDR data.*....Source: NTSB

161

approximately 1,200 fpm. Four seconds later, as the aircraft was about 110 feet agl and the captain stated, "I'll get over there," the vertical speed was computed by the NTSB as roughly 1,500 fpm.

FDR data indicated that an increase in EPR and nose-up elevator position began at about 4:38:32 P.M., and a second later the first officer stated, "nose up...nose up." The aircraft was about seventy-five feet agl, and the NTSB computed that the aircraft was descending at 1,800 fpm.

Two seconds later, the aircraft descent rate began to shallow. But the sink rate reduction would be too little, too late. The GPWS twice sounded its "sink rate" warning, and at 4:38:36.5 P.M., the G-limit switch on the FDR activated, indicating the moment of impact.

Captain's use of monovision contact lenses

The NTSB learned that about 1991 the captain began using monovision (MV) contact lenses to correct his near vision to 20/20. With MV contact lenses, a person wears a contact lens in one eye to correct for near vision, and another contact lens in the other eye to correct for distant vision. Traditionally both near and distant vision correction could be achieved by a person's use of bifocal glasses. More recently, however, the use of MV contacts has gained popularity.

The captain's optometrist told investigators that in general, "someone in a position of public safety" should use bifocal glasses rather than MV contact lenses when both near and distant vision correction was needed. He further stated that binocular vision correction (as opposed to monocular vision correction) would be preferable for pilots while performing flying duties, because there is a need for stable near and distant vision in cockpit situations. He reported that a pilot's use of MV contact lenses could impair sink rate perception, depth perception at some distances, and scanning vision, especially on the side with near vision correction. Investigators highlighted that the captain of Delta 554 was wearing the near vision correction lens in his left eye at the time of the accident.

The optometrist stated that MV contact lenses can impair depth perception, especially at distances of less than twenty to twenty-five feet. He indicated that this impairment could make it more difficult to land a small airplane or parallel park a car, but in his opinion, use of MV contacts should not adversely affect depth perception at greater distances.

The captain told investigators that he used MV contact lenses approximately seventy-five percent of the time that he flew. He indicated that he easily became accustomed to the MV contacts, and that he had not perceived any deficiency in vision or depth perception when wearing them. He stated he had not noted any visual problems wearing MV contacts while driving or flying.

According to the FAA's September 1996 *Guide for Aviation Medical Examiners,* "The use of a contact lens in one eye for distant visual acuity and another in the other eye for near visual acuity [monovision contact lenses] is not acceptable." The acting manager of the FAA's aeromedical certification division explained to investigators that this type of arrangement causes the pilot to alternate between one eye at a time, while effectively suppressing the other, causing a loss of binocular vision.

The captain told the NTSB that he was unaware of the prohibition against wearing MV contact lenses while flying. The aviation medical examiner (AME) that the captain used for FAA physicals since 1984 told investigators that although he was not specifically aware that MV contact lenses were not approved for use while flying, if the issue had come up, he would have advised airman medical certificate applicants not to use them while flying. He indicated that he did not recall discussing MV contact lenses with the captain, and he was unaware that the captain possessed any contact lenses, let alone MV contacts.

Investigators noted that AMEs are not required to ask pilots whether they possess contact lenses. They further noted that the FAA airman medical certificate application contains numerous questions concerning the applicant's medical history, but it does not ask whether the applicant wears contact lenses.

The NTSB interviewed several optometrists during the investigation and learned of no published information warning them that pilots were prohibited from using MV contact lenses while flying. Investigators cited a sales pamphlet provided by one MV contact lens manufacturer: "Monovision is a contact lens fitting technique that lets you see clearly both near and far...without the bother of bifocals." The pamphlet did not refer to any hazards or limitations associated with their use.

MV contact lenses cited in another accident
During the investigation, the Safety Board became aware of a previous general aviation accident involving a private pilot attempting to

land during degraded visibility due to low clouds and snow. In that mishap, the pilot later told investigators that he "didn't pull the nose up" during landing. The aircraft bounced hard on the runway and subsequently landed on the aircraft nose wheel, resulting in failure of the nose wheel tire. The Safety Board learned that the pilot had been wearing MV contact lenses when the accident occurred.

According to a flight instructor who was hired to fly with that pilot after that mishap, the pilot had a consistent tendency to delay the landing flare until too close to the ground. Coincidentally, the pilot returned to his AME to renew his medical certificate, where the AME noticed the pilot's use of MV contact lenses. He informed the pilot that MV contacts were not approved for flying. The flight instructor later flew with this pilot, this time without him wearing the MV contacts, and noticed that the pilot's landings "improved suddenly and dramatically" when he switched back to wearing glasses.

Visual cues and illusions

The NTSB cited the FAA's *Aeronautical Information Manual* (AIM) regarding how visual illusions can lead to spatial disorientation and landing errors. The AIM states that "landing over water, darkened areas, and terrain made featureless by snow, can create the illusion that the aircraft is at a higher altitude than it actually is. The pilot who does not recognize this illusion will fly a lower approach."

The AIM also states that atmospheric illusions, such as rain on the windshield, "can create the illusion of greater height" and that a pilot who does not recognize these illusions will also fly a lower approach. "Penetration of fog can create the illusion of pitching up. The pilot who does not recognize this illusion will steepen the approach, often quite abruptly." The AIM further explains that a pilot who is "overflying terrain which has few lights to provide height cues may make a lower than normal approach."

The AIM suggests that in order to mitigate consequences of these illusions, pilots should "use electronic glide slope or VASI systems." However, in the case of Delta 554, the electronic glide slope to Runway 13 was not usable below two hundred feet.

Conclusions

The NTSB found that the pilots held the appropriate flight and medical certificates, and they were trained in accordance with Federal

Aviation Regulations. However, they noted that the captain was using monovision (MV) contact lenses, which were not approved by the FAA for use by pilots while flying.

They further determined the aircraft was properly certified and maintained, and there were no air traffic control factors that contributed to the accident. For reasons cited in the "weather" section, above, the NTSB concluded that Delta 554 did not encounter wind shear during its approach and attempted landing. The investigation turned to the events within the cockpit for explanation of the accident's cause.

Delta 554's final approach

The NTSB noted that during the early portions of the approach, Delta 554 was stabilized in airspeed and descent rate, established on the electronic glide slope and in the landing configuration. However, when the TWA crew reported to the tower that they would be rejecting their takeoff, the captain of Delta 554 became concerned that he would have to execute a missed approach because of the possibility of the TWA jet still being on the runway. At this point he disconnected the autopilot and allowed the aircraft's descent rate to slightly decrease, flying above the electronic glide slope. The NTSB concluded that this reduction in descent rate was likely due to the "captain's anticipation of (and preparation for) a missed approach."

By the time the aircraft had deviated more than one dot above the glide slope, "it appears that the captain had recognized the deviation and had applied correction in an attempt to reestablish the airplane on the glide slope," stated the NTSB. The FDR showed that the engine thrust was reduced and the descent rate increased slightly. The NTSB stated that from 4:38:14 to 4:38:26 P.M. the airspeed and descent rate were generally steady and on target, and the airplane's position was such that a successful landing could be made. The Safety Board determined that at 4:38:21 P.M., as the aircraft descended through 200 feet agl, the Delta crew was in degraded visual conditions. Knowing that the electronic glide slope was unusable, they relied on other cockpit instrumentation and outside visual references for glidepath information. "The Safety Board concludes that because the airplane was in stable flight and the captain had taken actions to correct for a glide slope deviation, the captain's continuation of the approach after he established visual contact with the approach lights was not inappropriate."

FDR data indicated that about ten seconds before impact, engine thrust was reduced gradually. The Safety Board concluded that this was because the captain "perceived a need to slightly increase the airplane's rate of descent; however, the descent rate increased beyond what the captain likely intended to command." The captain then recognized that the descent rate required prompt attention and he was increasing aircraft pitch and engine thrust. About two seconds before impact, the aircraft was descending at 1,800 fpm, but the trend in vertical velocity had started to reverse. By the time the first officer called "nose up...nose up" the captain had already added power and increased the pitch; however, at this point, it was too late to avoid impact.

Visual factors

The Safety Board sought to determine why Delta 554's descent rate continued to increase near the end of the approach. They analyzed the visual cues in the airport environment, including the airport lighting system and the effect of the weather at the time of the accident, and the effect of the captain's vision limitations.

The NTSB noted that the pilots were in moderate-to-heavy rain and fog as they descended out of the clouds, and were flying over the waters of Flushing Bay, with no visible structures to aid in visually judging distance and/or altitude. The NTSB observed that the crew was facing all of the factors mentioned in the *Aeronautical Information Manual* (AIM) as contributing to visual illusions, including an absence of ground features (as when landing over water), rain on the windshield, atmospheric haze/fog, and terrain with few lights to provide height cues.

The NTSB further noted that the irregular and shorter spacing of Runway 13's runway edge lights could also have been a factor. "Pilots who are accustomed to operating into airports at which runway lights are spaced at constant 200-foot intervals might perceive their distance and angle to the runway differently when presented with runway lights spaced at shorter, irregular intervals," they stated. "The Safety Board concludes that the irregular and shortened runway edge light spacing and degraded runway weather conditions can result in a pilot making an unnecessary rapid descent and possibly descending too soon, especially in the absence of other visual references or cues."

The NTSB acknowledged that other aircraft were landing on Runway 13 at around the same time as Delta 554's attempt, and these aircraft also faced the same visual limitations and irregular runway light spacing; however, these other aircraft were able to operate without difficulty. The NTSB therefore evaluated how the captain's MV contact lenses could have effected his vision under those conditions. After considering all factors the NTSB concluded that "the captain's use of MV contact lenses resulted in his (unrecognized) degraded depth perception, and thus his increased dependence on monocular cues (instead of normal three-dimensional vision) to perceive distance." The Board remarked that because of the degraded visual conditions, "the captain was not presented with adequate monocular cues to enable him to accurately perceive the airplane's altitude and distance from the runway during the visual portion of the approach and landing. This resulted in the captain's failure (during the last ten seconds of the approach) to either properly adjust the airplane's glidepath or to determine that the approach was unstable and execute a missed approach."

"The Safety Board concludes that because of the captain's use of MV contact lenses, he was unable to overcome the visual illusions resulting from the approach over water in limited light conditions (absence of visible ground features), the irregular spacing of the runway edge lights at shorter-than-usual intervals, the rain, and the fog, and that these illusions led the captain to perceive that the airplane was higher than it was during the visual portion of the approach, and thus, to his unnecessary steepening the approach during the final ten seconds before impact."

The probable cause

The NTSB determined that the probable cause of this accident was "the inability of the captain, because of his monovision contact lenses, to overcome his misperception of the airplane's position relative to the runway during the visual portion of the approach. This misperception occurred because of visual illusions produced by the approach over water in limited light conditions, the absence of visible ground features, the rain and fog, and the irregular spacing of the runway lights."

"Contributing to the accident was the lack of instantaneous vertical speed information available to the pilot not flying, and the incomplete guidance available to optometrists, aviation medical examiners, and pilots regarding the prescription of unapproved monovision contact lenses for use by pilots."

Recommendations

Because of this accident, the NTSB made twelve recommendations to the FAA. Among these were:

- Identify Part 139 airports that have irregular runway light spacing, evaluate the potential hazards of such irregular spacing, and determine if standardizing runway light spacing is warranted. (A-97-84)

- Require all Part 121 and 135 operators to ensure that company manuals clearly delineate flightcrew (pilot flying/pilot not flying) duties and responsibilities for various phases of flight, and to ensure that "stabilized approach" criteria are clearly spelled out. (A-97-85)

- Revise the FAA Form 8500-8, "Application for Airmen Medical Certificate," to elicit information regarding contact lens use by the pilot/applicant. (A-97-86)

- Require that FAA's Civil Aeromedical Institute (CAMI) publish and disseminate a brochure to explain the potential hazards of MV contact lenses, and to emphasize to pilots that MV contacts are not approved while flying. (A-97-87)

- Require that all Part 121 and 135 operators notify their pilots and medical personnel of the circumstances of this accident, and to alert them to the hazards of MV contact lenses while flying. (A-97-88)

- Require all FAA Flight Standards District Office (FSDO) air safety inspectors and accident prevention specialists to inform general aviation pilots of the circumstances of this accident, and to alert them to the hazards of MV contact lenses while flying. (A-97-89)

- Require Part 121 and 135 air carriers to make their pilots aware of the type of vertical speed information that is displayed on their VSI (instantaneous/noninstantaneous), and to make them aware of the ramifications that type of information could have on their perception of their flight situation. (A-97-90)

- Require all Part 121 and 135 operators to convert noninstantaneous VSIs to instantaneous VSIs, where practical. (A-97-91)

- Require all Part 121 and 135 operators to review their flight attendant training programs and emphasize the need for flight attendants to aggressively initiate their evacuation procedures when an evacuation order has been given. (A-97-95)

The NTSB also recommended that optometric associations issue a briefing bulletin to member optometrists, informing them of the potential hazards of and prohibition against MV contact lenses by pilots while performing flying duties, and to urge them to advise pilot-rated patients of those potential hazards (MV contact lens's effect on distance judgments/perceptions). (A-97-96)

Industry action

An FAA test plan to determine the effects of variation of runway light spacing was completed in October of 1998. The full study is scheduled to be finished and the results published late in 1999. The Administration also issued a Flight Standards Handbook Bulletin (FSHB) 98-22, "Stabilized Approaches," to all principal operations inspectors (POIs). The bulletin will ensure that all operators' training and operations manuals contain appropriate criteria for the stabilized approach as referenced in the Operations Inspector's Handbook. That manual also details each pilot's flying and nonflying responsibilities during the approach phase of flight.

To educate both AMEs and the flying public, the Civil Aeromedical Institute (CAMI) updated its *Medical Facts on Pilots/Pilot Vision* publication, and an editorial was included in the Federal Air Surgeon's Medical Bulletin regarding nonstandard vision approaches. Furthermore, revisions to the *Airman's Information Manual* (AIM) and the FAA Medical Handbook for Pilots are planned to include warnings about monovision lenses. The FAA notified all flight standards regional managers and aviation safety inspectors of the hazards of monovision lenses, and will revise Form 8500-8, "Application for Airman Medical Certificate," to include a question to determine contact lens use by the applicant.

Early in 1998, two Flight Standards Information Bulletins (FSIB) were issued. The first, "Vertical Speed Indicator Knowledge Needed by Pilots," directs all POIs to assure that air carriers inform their pilots of the differences between instantaneous and noninstantaneous indicators. A possible Notice of Proposed Rulemaking (NPRM) to convert all indicators to the instantaneous type is under review. The second FSIB, "Need for Flight Attendants to be Aggressive in Initiating Emergency

Evacuations," emphasizes the importance of rapid initiation of evacuation procedures when an evacuation order has been given

Epilogue

Those traveling aboard Delta 554 were lucky. They narrowly missed being involved in a major aviation disaster—the word "narrowly" being defined here literally in terms of inches. The aircraft suffered some $14 million in damage. But airplanes can be repaired; human lives cannot. In this accident, fortunately, there was no loss of human life. Not even a serious injury. The lessons learned were cheap by any standard. But it remains to be seen just how far the industry will go to prevent this type of accident in the future.

References and additional reading

Federal Aviation Administration (FAA). 1999. *NTSB Recommendations to FAA and FAA Responses Report.* <http://nasdac.faa.gov/>

Flight Safety Foundation. 1997. MD-88 strikes approach light structure in nonfatal accident. *Accident Prevention.* December, 1-16.

National Transportation Safety Board (NTSB). 1997. *Aircraft accident report: Descent below glidepath and collision with terrain. Delta Air Lines flight 554. McDonnell Douglas MD-88, N914DL. LaGuardia Airport, New York. October 19, 1996.* NTSB/AAR-97/03. Washington, D.C.: NTSB.

8

A final evaluation flight of Airborne Express

Operator: ABX Air (Airborne Express)

Aircraft Type: Douglas DC-8-63

Location: Narrows, Virginia

Date: December 22, 1996

Airline operational safety is based in part upon effective flight train-ing, experienced crews, and standardized, routine procedures. Ad-ditionally, detailed guidelines for all phases of flight operations are available to pilots in the form of comprehensive flight manuals. Used together, these components combine to produce an extremely safe operational environment. If any one of these ingredients is substan-dard, the margin of safety is degraded—the loss of all of them at once, however, is a sure recipe for disaster.

Flight history and background

N827AX, a DC-8-63 (see Fig. 8-1), was built in 1967 and had accu-mulated over 63,000 hours of flight time. It had flown for several ma-jor airlines before being completely reconditioned for use by the small package cargo airline Airborne Express (ABX). Triad Interna-tional Maintenance Corporation (TIMCO) of Greensboro, North Car-olina (GSO), had completed a "D" check, or major overhaul of the airframe and engines, and had installed new navigation and com-munication gear, radar, electronic flight instrumentation, cargo-han-dling equipment, and Stage III noise suppression systems.

Originally scheduled for revenue service in late October of 1996, the modifications and other unplanned maintenance tasks delayed delivery of N827AX by two months. But ABX had already contracted to operate charter flights based upon the availability of this aircraft, so any further delay was unacceptable. Once a successful final evaluation flight (FEF) was accomplished, the airplane would immediately join the thirty-five other DC-8s in the ABX fleet.

The scheduled FEF was to be a very comprehensive test flight. All aircraft systems would be activated and tolerances checked, landing gear and wing flaps cycled, indicators calibrated, and flight-handling characteristics verified. A complete stall series would also be conducted to confirm that neither aircraft modifications nor rerigging of the control cables had altered expected aerodynamic performance. Management pilots from the flight standards departments normally comprised the flightcrews for the FEFs, as they usually had flight check experience and the required technical expertise.

In anticipation of the FEF, the crew arrived in GSO early in December of 1996. The pilot in command was the former manager of DC-8 flight standards for ABX, who had been promoted to the same position for the new B767 fleet. He would occupy the right seat in the cockpit, acting as an instructor for the new DC-8 manager of flight standards who would fly the airplane from the left seat. The flight engineer was a flight standards flight engineer, and all three crewmembers were FAA-designated DC-8 examiners. Also on board were two ABX maintenance technicians and a TIMCO employee to assist with in-flight system evaluations.

After completing all the necessary paperwork and preflight inspections, the FEF was scheduled for December 21, 1996. Once airborne, however, a minor hydraulic malfunction forced an early termination of the flight and return to GSO for repairs. Originally scheduled to resume the flight at 1:30 P.M. (EST) the next day, it was 5:40 P.M. before the aircraft again took off. Instead of a daytime operation as initially planned, the entire flight would now have to be flown at night.

An unexpected landing gear warning horn sounded just after lift off, so after climbing through 4,000 feet mean sea level (msl) the landing gear was extended and retracted. No additional warnings were heard. ABX 827 was then assigned a "block" altitude clearance by the Atlanta Air Route Traffic Control Center (ARTCC) which allowed them to operate at any altitude between 13,000 and 15,000 feet. Leveling off at 14,000 feet, the gear was again cycled, but this time with

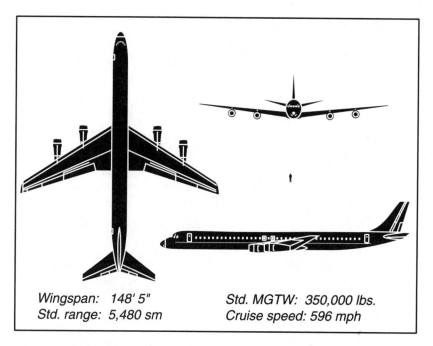

Wingspan: 148' 5"
Std. range: 5,480 sm

Std. MGTW: 350,000 lbs.
Cruise speed: 596 mph

8-1 *Douglas DC-8-63.*

the flaps extended. Again the warning horn sounded. "Little weird," commented the pilot-not-flying (PNF) in the right seat.

A test of the engine anti-icing system was done and one component was found inoperative. Vectors were issued by the ARTCC for traffic avoidance, and a series of hydraulic systems tests were begun. At 5:48 P.M., the pilot flying (PF) stated "we're getting a little bit of ice here, we...probably get out of this." The aircraft was skimming through the tops of the clouds, and four minutes later he said, "We just flew out of it. Let's stay here for a second."

A check of flap extension times and elevator trim operation was done, but a master caution light momentarily diverted the crew's attention. An erroneous manifold high temperature indication was resolved and the tests continued.

The flight engineer stated "OK, flaps fifty at one point three Vs zero, to flaps twenty three while accelerating to one point five Vs twenty three. It should be both engine hydraulic pumps on to give me a chance to get configured."

Again the master warning light illuminated with an overtemp indication. A quick check verified acceptable system status, and the litany continued. "OK, flaps twenty three."

"There's one sixty," announced the PF.

"OK, want to try it again?" replied the PNF.

"Yeah, let's go...back to fifty."

"Back to fifty and slow..."

The flaps were repeatedly cycled with normal and alternate hydraulic power, and all test results were recorded. The flap position indicator was noted as being slightly inaccurate.

"OK, next thing," called the flight engineer. "Standby rudder pump operation. Aileron power lever on. Both engine hydraulic pumps bypassed. Standby rudder pump on...aileron power lever, ready for it to come off?"

And so it continued for the next several minutes. At 6:05 P.M., twenty-five minutes after take off, the engineer informed the pilots that the next required maneuver was the stall series. The PNF stated "One eighty four and...we should get, uh stall at one twenty two. I'm gonna set that in my interior bug."

"Shaker at twenty eight. If you just call out your numbers, I'll record them," the engineer responded.

In order to reduce airspeed, the four JT-3D engines had to be throttled back. But acceleration from idle to any commanded thrust level takes several seconds, and at low power settings and airspeeds abnormal airflow through the engines can cause compressor stalls, which cause loud popping noises and can lead to engine compressor damage. Maintaining engine RPMs above idle (known as being "spooled") assures power will be instantly available for the stall recovery and prevents engine damage as well.

The PNF, acting as an instructor, reminded the pilot flying that "the only trick to this is just don't unspool."

"Unspool and then I'll respool," acknowledged the PF.

As the airspeed decreased below 173 knots the PNF stated, "Guess I better not trim below...two..." and seven seconds later, "Yeah, I'm gonna spool now." All four engines then increased in power.

At 149 knots, the PF called "Some buffet!" indicating the earliest signs of an aerodynamic stall. "Yeah, that's pretty early," responded another crewmember. The whole aircraft began to shake. "That's a stall right there! Ain't no shaker!" exclaimed the engineer. The PF called for max power to be set as he started the recovery procedure. Seven seconds later one engine experienced compressor stalls, recorded as loud banging noises on the cockpit voice recorder (CVR).

Twenty seconds after recognizing the stall, the PNF urged the PF to "Take a little altitude down, take it down!" Maximum power had been applied, but the PF was maintaining backpressure on the control column, so the aircraft was still decelerating. Just a few moments later the airplane was fully stalled and began a very fast, flat descent. Because the nose was still raised, the wing's angle of attack was very high and airspeed was still below stall speed. The airplane rolled hard to the left, past 75 degrees, then back to the right. Again and again the wings rolled until almost vertical, only to swing back rapidly the other direction. The PNF tried to talk the PF through the recovery. "A little rudder," and "start bringing the nose back up." ATC called to ask if ABX 827 was in an emergency descent, to which they replied, "yes sir." There would be no further radio transmissions from the aircraft.

The airplane was falling out of the sky at over 6,000 feet per minute, with wildly fluctuating airspeed indications. On the flight deck, the situation was getting desperate. The PF was still pulling hard on the control column trying to force the DC-8 to fly. Power was added, then removed. "Rudder, rudder!" yelled instructor. "Left rudder!"

"Left rudder's buried," was the reply.

"OK, easy, don't...OK now, easy, bring it back."

The aircraft had now rolled 113 degrees to the right, well past vertical. The nose was pitched down fifty-two degrees and still falling. The pilot had been struggling to regain control for a minute and a half, but still he wrestled with the controls, forcing the airplane to roll back to the left and the nose to rise slightly. But both time and altitude were running out. The Ground Proximity Warning System (GPWS) started to shriek "Terrain, terrain, whoop whoop, pull up!" and three seconds later N827AX slammed into a Virginia mountainside.

The wreckage path was only seven hundred feet long. Left wing low with a flight path angle of thirty-five degrees down, the airplane's speed was about 240 knots when it plowed into the forest. Some pieces of the leading edge were ripped from the wing, but the rest of the severely fragmented fuselage remained within the impact crater and small fan-shaped debris field. All aboard died instantly. (See Fig. 8-2.)

The investigation and findings

The National Transportation Safety Board (NTSB) dispatched investigators to the scene specializing in meteorology, maintenance, airline operations and aircraft structures, systems, and powerplants. Investigative groups also met in Washington, D.C., for review and transcription of the CVR and flight data recorder (FDR). Parties to the investigation included the FAA, Douglas Aircraft Co., Pratt & Whitney (the engine manufacturer), ABX Air, Inc., and TIMCO.

The investigation and subsequent analysis focused on the performance of the flightcrew, their training and prior experience, the organizational structure of ABX, and FAA oversight of the airline's operation.

Flightcrew factors

The Safety Board found that the flightcrew was qualified and properly certificated to perform the FEF.

8-2 *Wreckage of Airborne Express N827AX on a Virginia mountainside*...Courtesy FAA

Crew experience

A review of company records by the Safety Board revealed that the PNF had only about one hour of PIC time during an FEF, that being the day before the accident, while the flying pilot, occupying the left seat, had never before flown a DC-8 FEF. These flights are, by their very nature, nonroutine and require special operational considerations. The Safety Board found that ABX's training program was "informal and undocumented [and required no] specific training or proficiency requirements for pilots conducting FEFs." The NTSB reasoned that it was this informality that allowed two pilots who had never handled the controls during an actual stall in the DC-8 to be paired as a flightcrew on an FEF.

Both pilots had management duties that occupied a great deal of their time. In the year prior to the accident, the PF had accumulated only eighty-nine hours of flying time, all in the DC-8, while the PNF had flown only 115 hours, also exclusively in the DC-8.

ABX maintained an approved Crew Resource Management (CRM) training program, but by December of 1996 only about half of the line pilots had completed the two-day course. The PF had been an instructor in the program, but neither the engineer nor the PNF had received the mandatory training. No recurrent CRM training procedures had yet been developed by the airline for either classroom or simulator use.

The ABX DC-8 flight simulator

The FAA-approved simulator training conducted by ABX emphasized minimum altitude loss during the stall recovery. In this training, maximum power is always upon stall recognition, but pitch attitude is decreased only enough to gain adequate airspeed for the recovery. Avoidance of a secondary stall, maintaining aircraft heading, and minimum altitude loss are the criteria on which successful completion of the maneuver were based.

Pilots normally practice stall recovery techniques only in the airline's flight training simulator, not in the aircraft. As a training device for normal flight operating regimes, a simulator is a valuable tool. But it was never designed to reproduce aerodynamic characteristics of maneuvers that line pilots will not encounter, or those that are not taught (such as recovery from a fully developed stall). In evaluating the ABX DC-8 simulator, the NTSB found that its stall characteristics were much more benign than those of the actual aircraft. No stall break, or abrupt nose-down movement was evident, and once well below stall speed, the simulator entered a stable descent, exhibiting no aerodynamic buffet or continued decrease in airspeed. As a result, the Board was concerned that these differences may have confused the flightcrew when confronted with an actual stall recovery attempt, resulting in the inappropriate flight control inputs.

The stall recovery attempt

The Safety Board found that for the most part, the flightcrew prepared for the stall series properly by slowly reducing airspeed and maintaining altitude and engine RPM. Stall recognition, the decision to terminate the stall, and initiation of recovery were timely and in accordance with ABX procedures.

Normal recovery technique requires a relaxation of back-pressure on the control column, or even a slight forward movement to provide proper pitch control and adequate aircraft acceleration. But FDR analysis indicated that in this case aircraft pitch attitude was maintained between ten and fourteen degrees nose up for the first eight seconds after stall recognition, allowing the airplane to enter the fully developed stall. Furthermore, the control column was only moved aft, never forward, from five degrees to twenty degrees in the first twenty-two seconds of the event.

As the aircraft began a steeper descent, its pitch attitude never decreased. This resulted in an increasing angle of attack (angle of the wing to the relative wind) which deepened the stall. Without an aggressive forward movement of the control column to decrease the angle of attack, recovery was impossible. The FDR confirmed that the PF held the column aft all the way to ground impact.

According to the NTSB, the PNF should have noticed the full aft position of the control column, the continued aircraft buffeting, the extreme nose-down pitch, and the low airspeed/high rate of descent of the airplane. Although at one point he suggested that the PF "take it down," to trade altitude for airspeed in order to recover from the stall, he made no further effort to take control of the aircraft. The Safety Board was concerned that because both pilots were management captains, there may have been some ambiguity as to actual command authority.

Other factors affecting stall recovery

The NTSB investigated a number of other factors that may have influenced the performance of the flightcrew during the stall recovery, including the failure of the airplane's stall warning system, stall buffet at an earlier-than-expected airspeed, marginal weather conditions, and engine compressor surges.

The stall warning stickshaker never activated during the accident flight, even though preflight testing indicated a fully functional system. Total destruction of all stall warning components found at the accident site precluded the NTSB from determining why the stick shaker did not operate in flight. The Safety Board did determine that the failure of the system did not prevent the crew from recognizing the onset of the stall, but may have contributed to their confusion regarding airplane's accelerated stall condition as the angle of attack continued to increase.

Based on data supplied by Douglas Aircraft, the calculated stall speed for the accident aircraft was 122 knots, with an initial buffet at 137 knots. The CVR recorded the crew's comment of "some buffet" at 149 knots or about twelve knots earlier than expected. Three factors could have affected this speed: weight of the aircraft, airframe icing, and rigging of the flight controls and flaps.

According to the Safety Board's analysis of the accident airplane's takeoff performance and known loading schedules, all weight and balance computations were done correctly. However, a combination of icing encountered during the climbout and possible misrigging during overhaul may have resulted in a higher-than-anticipated buffet speed. The Board also determined the actual speed that the airplane stalled at was 126 knots, only four knots higher than expected.

The stall series was begun in darkness, immediately above the tops of a cloud layer. The Safety Board determined that no natural horizon would have been visible to the crew, and once the descent from 14,000 feet was started, instrument meteorological conditions (IMC) would have existed all the way to the ground. While the director of flight technical programs at ABX preferred that FEFs be conducted during daylight hours and in clear air, no official company policy or guidance was available to the flightcrew. The NTSB determined that a visible natural horizon would have greatly assisted the crew in determining their actual flight path, enabling avoidance of the stall completely or at the very least, a rapid stall recovery.

Finally, it was the Safety Board's opinion that loud engine compressor surges may have distracted the crew during a critical period of flight. These compressor stalls first occurred just as the airplane entered the aerodynamic stall, when proper lateral control inputs could have assured full recovery.

ABX FEF procedures

In May of 1991, the crew of another ABX DC-8-63 FEF lost control while executing a stall series, but with the landing gear and flaps extended. The stick shaker and the stall buffet occurred simultaneously, and two engines experienced severe compressor surges. A rapid roll and yaw began, and the aircraft entered a spin and lost 6,000 feet of altitude before the crew regained control.

Changes agreed to by the FAA's Principal Operations Inspector (POI) assigned to the airline and ABX personnel included mandatory FEF training in the simulator, the requirement for an expanded, 3,000-foot block altitude for stall testing, and a significant change in stall recovery procedure. The new policy was to use pitch (lowering the nose of the aircraft) first and then power (advancing the throttles to maximum) to fly out of the stall. But the ABX director of flight technical programs, who was on a leave of absence during this time, disagreed with the FAA—upon his return to active service he did not advocate nor use the revised procedures. The accident flight pilot in command (the pilot not flying) was originally trained in stall recovery techniques by the previously on-leave director, using the older "power only" method. He then taught the accident PF, again using only power, not pitch, for recovery.

The NTSB believed that the accident flightcrew was expecting no altitude loss during the maneuver, as they requested only a 2,000-foot altitude block and began the stall series only 500 feet above their minimum assigned altitude. The Safety Board found no evidence that ABX had officially incorporated into their FEF programs the stall recovery changes made as a result of lessons learned from the prior DC-8 loss of control incident.

FAA oversight

An FAA National Aviation Safety Inspection Program (NASIP) audit of ABX in 1991 found the airline to be in compliance with the Federal Aviation Regulations (FARs) but that some flight operations were being conducted without written corporate policy or guidance. Additionally, the check airman/instructor training records were inadequate and the training programs lacked appropriate lesson plans.

While there was no requirement for FAA surveillance of the FEF program, the same problems found in the NASIP applied to flightcrew training and qualifications for FEFs as well. The Safety Board determined that subsequent attempts to correct these problems addressed only a few operational deficiencies and failed to resolve the more serious safety issues pertaining to overall airline operations as identified by the inspection and the 1991 loss of control incident.

Conclusions and probable cause

Other significant findings by the Safety Board were that the accident may have been prevented if an angle-of-attack indicator had been available for use by the flightcrew, and that in general, there is a need for additional "procedural definition and training measures" in all nonroutine air carrier operations. The Board also found that "The currently established procedural requirements for conducting functional evaluation flights on large transport aircraft provide inadequate guidance to air carrier operators, maintenance repair stations, FAA principal operations and maintenance inspectors and other affected parties."

As stated in the final accident report, "The National Transportation Safety Board determines that the probable causes of this accident were the inappropriate control inputs applied by the flying pilot during a stall recovery attempt, the failure of the nonflying pilot-in-command to recognize, address, and correct these inappropriate control inputs, and the failure of ABX to establish a formal functional evaluation flight program that included adequate program guidelines, requirements, and pilot training for performance of these flights. Contributing to the causes of the accident were the inoperative stick shaker stall warning system and the ABX DC-8 flight training simulator's inadequate fidelity in reproducing the airplane's stall characteristics."

Recommendations

Following the fatal accident near Cali, Colombia, in 1995 (see Chapter 2), the NTSB published recommendation A-96-94 that would require angle-of-attack instrumentation to be available to the flightcrew so that aircraft climb performance could be maximized. The Safety Board determined that accident might have been avoided had the flight crew had this important instrumentation, and reiterated their support for this previously issued recommendation. The FAA, however, for reasons discussed in the "Industry action" section of Chapter 2, decided not to mandate those displays.

As a direct result of the ABX investigation, the NTSB issued seven new recommendations, requesting the FAA to:

- Review all stall warning system test and calibration procedures as found in the DC-8 maintenance manual to assure timely calibration and functional checks of all components of the system. (A-97-46)

- Evaluate the stall characteristics of each air transport aircraft and assure that the applicable flight training simulators reproduce those stall characteristics accurately. Once that is complete, add pilot training in the recovery from low pitch attitude stalls to the "special events" training program of air carriers. (A-97-47)

- Ensure that ABX incorporates the specific revised stall recovery technique (as agreed to in 1991) into their DC-8 FEF program. (A-97-48)

- Produce an advisory circular (AC) for distribution to air carriers that provides guidance in all phases of FEF operations, and modify the Federal Aviation Regulations (FARs) to reflect those procedural requirements. (A-97-49 and 52)

- Identify any special operation that might be conducted by an air carrier, including FEFs or other nonroutine flights, and assure that the carrier's operations specifications contain appropriate guidelines and limitations for conducting those flights. Amend the applicable FARs to mandate special flightcrew training and qualification requirements for these flights within each air carrier's training manual. Once that has been accomplished, specifically monitor the airline's FEF program. (A-97-50 and 51)

Industry action

The FAA and Douglas Aircraft completed a comprehensive review of the DC-8 stall warning system maintenance practices and found existing procedures thorough and appropriate. No changes are planned.

The National Simulator Program staff of the FAA evaluated the possibility of programming air transport simulators to accurately repro-

duce in-flight handling characteristics during stalls or other maneuvers outside the normal operating environment of these aircraft.
They discovered that accurate flight performance data is not readily
available, would be very costly and dangerous to obtain, and would
be of questionable accuracy. There are analytical methods that could
be employed, but supporting flight test data would be necessary for
accuracy and reliability. The FAA plans no further action on this
item. However, an industry task force was convened to evaluate current pilot training and checking procedures in stall maneuvers. That
group determined that the current training requirements "place un-
due emphasis on the importance of minimum loss of altitude during
stall recovery." A revision to 14CFR 121 Subparts N and O will reflect
new requirements as they are developed, ones that will no longer insist on minimum loss of altitude in all stall recovery conditions. The
FAA has also revised its Practical Test Standards to reflect its new
thinking, stating that altitude loss during stall recovery must be kept
to a minimum. But "...at intermediate and higher altitudes, airspeed
and/or altitude loss not necessary for the safe and expedient recovery from the approach-to-stall condition should be avoided."

In November of 1997, ABX fully incorporated all previously agreed-
to stall recovery techniques into their FEF procedures. A new FEF
training program conducted in the simulator emphasizes those maneuvers, including stalls, not routinely encountered in normal line
operations.

After reviewing Parts 61 and 91 of the FARs, the FAA has determined
that regulatory changes affecting FEF flightcrew training and qualification requirements are warranted at this time. Since all FEFs must
be conducted under Part 91, guidance should be provided by a flight
standards handbook change, which would ultimately be incorporated into the FAA's operations and airworthiness inspectors' hand-
books. The FAA believes that working through each airline's
principal inspectors provides the most expedient way to address this
problem.

And, although not in direct response to this accident, the aviation industry has brought new focus to loss of control accidents. After analyzing accidents over a thirty-year period, the Safety Board in 1996
issued recommendation A-96-120, requiring air carriers to provide
training to flightcrews in the recognition of and recovery from any
unusual attitude, whether caused by flight control malfunctions or

uncommanded control surface movement. Training should also include upsets while the aircraft is being controlled by automatic flight control systems. The FAA is considering a Notice of Proposed Rulemaking (NPRM) that would require airlines to conduct training that would "emphasize recognition, prevention and recovery from aircraft attitudes not normally associated with air carrier flight operations." One industry effort, supported by thirty-four airlines, professional associations, and manufacturers, developed a comprehensive training aid, titled "Airplane Upset Recovery," which was published by Airbus Industrie in December of 1998.

The bottom line, as the FAA recently stated, is that in a stall recovery situation, "the first order of business is to do whatever is necessary to recover—to regain the ability to 'fly'."

Epilogue

This accident, although a very real tragedy to the families of the crewmembers, didn't make the national news. It involved a nonrevenue, cargo airline flight and an older, out-of-production aircraft that crashed in a remote area. The NTSB found the pilots to be at fault, thus assuring a low profile conclusion to an unfortunate mishap. But real, relevant, and timely lessons were learned, ones that can be applied to all aviation operations.

The Safety Board has rightfully pointed out that nonroutine operations are demanding upon both the crews and their aircraft. These flights highlight the importance of diligent maintenance and exceptional flightcrew training and qualifications. Appropriate guidance to operators is essential, lessons learned from previous incidents must be wholly embraced, and consistency of operational considerations throughout the industry is a necessity. But it really doesn't matter if it's an FEF or the afternoon "milk run" to Peoria—practicing these hard-won principles makes the skies safer for everyone.

References and additional reading

Aarons, Richard N. 1997. Cause and circumstance: Post-maintenance flight test hazards. *Business & Commercial Aviation*. May, 98.

Federal Aviation Administration. 1998. *NTSB Recommendations to FAA and FAA Responses Report*. <http://nasdac.faa.gov/>

Flight Safety Foundation. 1997. After intentionally stalling DC-8, crew uses incorrect recovery technique, resulting in uncontrolled descent and collision with terrain. *Accident Prevention.* September.

Garrison, Peter. 1998. Aftermath: Unrecognized stall. *Flying.* January, 81-83.

National Transportation Safety Board. 1997. *Aircraft Accident Report: Uncontrolled flight into terrain. ABX Air (Airborne Express), Douglas DC-8-63, N827AX, Narrows, Virginia, December 22, 1996. NTSB/AAR-97/05.* Washington, D.C.: NTSB.

National Transportation Safety Board. 1997. Public docket. *ABX Air (Airborne Express), Douglas DC-8-63, N827AX, Narrows, Virginia, December 22, 1996. DCA97MA016.* Washington, D.C.: NTSB.

Phillips, Edward H. 1997. NTSB faults pilots in fatal DC-8 crash. *Aviation Week & Space Technology.* 21 July, 42.

9

Waiting for rescue

An agonizing night for KAL 801

Operator: Korean Air, Inc.

Aircraft Type: Boeing B-747-300

Location: Agana, Guam

Date: August 6, 1997

Sometimes the lessons we learn from aviation accidents are, unfortunately, repetitive. A few of the previous chapters and a few later in this book discuss "Controlled Flight Into Terrain (CFIT)," presently the leading cause of aircraft accidents and passenger deaths worldwide. In some respects, KAL 801 will give the aviation safety industry yet another opportunity to delve into the many facets of this perplexing topic. But in one awful, appalling way, this accident stands out from all the others.

Flight history and background

She was the undisputed queen of the sky. As tall as a six-story building and more than 225 feet long, the B-747-300 was the largest commercial passenger jet in the world (see Fig. 9-1). Three flight deck crewmembers, a purser, and thirteen flight attendants were assigned to Korean Air (KAL) flight 801 that Tuesday night to provide the 260 passengers on board a comfortable journey.

Earlier that evening, the flightcrew had met in the KAL dispatch center to review the flight release and weather, collect all required paperwork, and conduct the various briefings. At 9:27 P.M., flight 801 departed Kimpo Airport in Seoul, Korea, after a short ground delay, for the three-hour-and-fifty-minute trip to Guam. Because the Cockpit Voice Recorder (CVR) was designed to record only the final thirty minutes of flight, our knowledge of cockpit conversation is limited to that time frame. At the beginning of the recording, cockpit preparations had already begun for the imminent descent and landing at Guam International Airport (PGUM) in Agana, Guam.

"...It was a little lengthy, but that completes my landing briefing," concluded the captain. Having previously received the destination weather, he confirmed that the reported visibility of six miles would allow a visual approach. Reviewing the Instrument Landing System (ILS) Runway 6 L (left) approach plate, he confirmed the current altimeter setting, desired descent point, and aircraft approach speeds. Aware that the glideslope (electronic precision vertical guidance information) portion of the ILS was inoperative, he acknowledged an MDA (minimum descent altitude) of 560 feet mean sea level (msl), with a corresponding HAT (height above touchdown) of 304 feet above ground level (agl). The missed approach procedure, while not anticipated, was carefully reviewed.

Guam Center and Radar Approach Control (CERAP) had previously issued a clearance to 2,600 feet, so a gradual descent was begun out of FL 410 (41,000 feet). Three minutes later, the captain complained to the rest of the crew about their demanding flight schedule.

"If this round trip is more than a nine-hour trip, we might get a little something...they [KAL] work us to maximum, up to maximum...Probably this way," he continued, "hotel expenses will be saved for cabin crews, and they maximize the flight hours. Anyway, they make us [B-747] classic guys work to maximum." The aircraft and crew were scheduled to return to Seoul very early the next morning, after a scheduled ground time of only three-and-a-half hours. Typically, crews would nap on the aircraft during the break.

"Really... sleepy," mumbled the captain. The first officer concurred.

Both pilots were monitoring the on-board weather radar, which was indicating widespread areas of rain over the island. "Captain, Guam condition is no good," advised the first officer.

"It rains a lot," confirmed the captain. "Request twenty-mile deviation later on... to the left as we are descending."

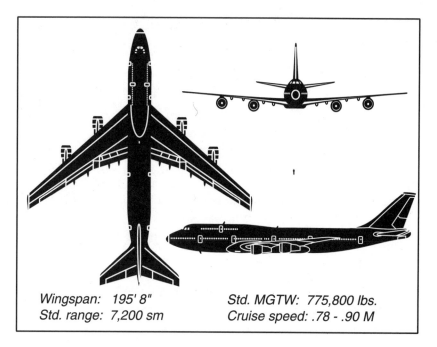

| Wingspan: 195' 8" | Std. MGTW: 775,800 lbs. |
| Std. range: 7,200 sm | Cruise speed: .78 - .90 M |

9-1 *Boeing 747-300.*

The first officer then pointed out weather along that route. "Don't you think it rains more in this area here?" he asked.

"Left, request deviation," ordered the captain. "One zero miles!"

"Yes," acknowledged the first officer, acquiescing to the senior pilot's instructions.

All three flight deck crewmembers had served in the Korean Air Force, but the captain, only two years older than the first officer, had been with the airline almost ten years while the first officer had only three-and-a-half years with the company. The flight engineer was eighteen years older than either of the pilots and had been with KAL the longest, at over eighteen years. A navigator in the Korean Air Force, he had never been a pilot.

The descent checklist was completed and a further discussion ensued regarding the thunderstorms and the flight's course deviations. "Today the radar has helped us a lot!" offered the flight engineer. It had not been a smooth flight—moderate turbulence encountered earlier had been uncomfortable, requiring an interruption of the in-flight meal service.

"Yes, it is very useful," replied the captain. "Request heading one sixty." CERAP approved the heading.

Within a few minutes, the airliner was clear of the rain shower, and the crew requested a radar vector from the controller to Runway 6L at Agana. The approach checklist was read and the ILS frequency was tuned and identified on the number one navigation radio. The "reminder bugs" on the radio altimeters were set to 304 feet, as required by the localizer only (glideslope out) approach plate. The radar detected yet another thunderstorm, this one to the left of their flight path, and light turbulence jostled the airplane.

"Flaps one," commanded the captain.

"Flaps one, one nine nine," answered the first officer, acknowledging the flap selection and stating the minimum maneuvering speed for that flap configuration and aircraft weight.

"Five," called the captain, as the aircraft continued its descent.

"Flaps five, one seventy nine."

The altitude alerter sounded, reminding the crew that they were only a thousand feet above their assigned altitude. "Flaps ten."

"Korean Air eight zero one turn left heading zero niner zero, join localizer," radioed the controller.

"Heading zero nine zero, intercept the localizer," confirmed the first officer. The turbulence was subsiding, but an occasional wind gust rocked the massive airplane.

"Oooh!" remarked a surprised crewmember as the aircraft bounced through a rain shower. "Cool and refreshing!"

"Glideslope...localizer capture," announced the first officer. By then their position was north of, but abeam the FLAKE intersection (see Fig. 9-2), and level at 2,600 feet.

"Korean air eight zero one, cleared for ILS Runway six left approach...glideslope unusable," reminded the combined center/approach controller.

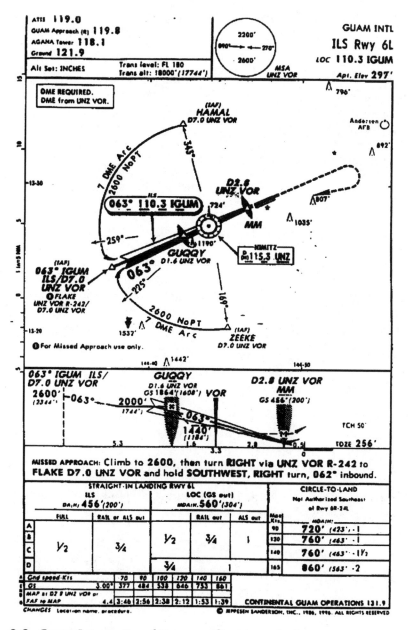

9-2 *Guam International Runway 6L approach chart as reproduced in the public docket...* Source: NTSB

"Korean eight zero one roger...cleared ILS Runway six left," was the reply. No acknowledgement was made of the inoperative glideslope.

On the flight deck, some movement was detected in the ILS display's vertical guidance pointer, causing the flight engineer to question, "Is the glideslope working? Glideslope? Yeah?"

"Yes, yes, it's working!" exclaimed the captain.

"Ah, so..." agreed the flight engineer.

One crewmember asked, "Check the glideslope if working? Why is it working?"

The first officer replied, "Not usable!"

"Six D check, gear down." The altitude alerter sounded again and the aircraft continued to sway in the light turbulence. They were then two miles inside of FLAKE intersection, and just over eight miles from the runway. A descent out of 2,600 feet was started even though confusion still prevailed as to the status of the ILS.

"Glideslope is incorrect." The descent continued.

"Approaching fourteen hundred," called the first officer.

Four seconds later, the captain stated, "Since today's condition of the glideslope is not good, we need to maintain one thousand, four hundred forty. Please set it." No mention was made of the intermediate level-off altitude of 2,000 feet msl as depicted on the approach chart.

"Korean air eight zero one, contact the Agana tower, one one eight point one. Ahn nyung hee ga sae yo [goodbye in Korean]."

"Soo go ha sip si yo [take care in Korean]," responded the first officer to the controller, "one eighteen one." The flight engineer commented that the controller had probably been a "GI" once stationed in Korea.

Passing through 2,100 feet, sounds of the aircraft configuration warning horn and altitude alerter were heard over the cockpit speakers (see Fig. 9-3). "Agana tower, Korean air eight zero one intercept the localizer six left," radioed the first officer.

"Korean air eight zero one heavy, Agana tower, Runway six left wind...at zero niner zero at seven...cleared to land," responded the lone local controller in the tower cab. "Verify heavy Boeing seven four seven tonight." Since the flight was usually flown by one of the airline's Airbuses, verification of aircraft type was necessary.

"Korean eight zero one roger...cleared to land six left," answered the first officer. The CVR captured the sound of flap handle movement at about the same time.

"Flaps thirty," commanded the captain.

"Flaps thirty...flaps thirty confirmed," announced the first officer. The configuration warning horn sounded again, and the aircraft continued its descent through 1,500 feet. "Landing check?"

Still one-half mile outside the GUQQY outer marker and over five miles from the runway, KAL 801 descended through 1,400 feet msl. Anticipating a visual sighting of the airport, the captain reminded the crew to, "...Look carefully. Set five hundred sixty feet [in the altitude alerter, published MDA for the approach]."

"Set," confirmed the first officer. The landing checklist was begun.

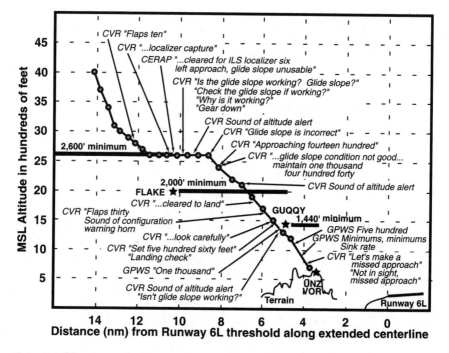

9-3 *Profile view of KAL 801 radar data and selected communication excerpts. . . .* Source: NTSB

The digitized voice of the Ground Proximity Warning System called its first alert as the airplane was passing through 1,400 feet. "One thousand," it warned, as the radio altimeter indicated 1,000 feet above the undulating terrain. The checklist continued.

As the altitude alerter chimed once more, the captain again asked, "Isn't glideslope working?" The rain increased, prompting a command for the flight engineer to activate the windshield wipers. The B747 continued to descend, crossing the GUQQY marker at about 1,200 feet, well below the 2,000 feet procedurally required.

The captain was becoming concerned that the runway was not yet visible. "Not in sight?" he asked of the other crewmembers, just as the GPWS intoned "five hundred."

"Eh?!" was his instinctive, astonished response to the warning. They were still almost five miles from the runway.

"Stabilize, stabilize!" cautioned the first officer. .

"Oh, yes," he answered. But the aircraft did not level off, instead continuing to sink at a rate even greater than before. The flight engineer continued reading the checklist. Ten seconds later the GPWS blared, "minimums, minimums!" The aircraft was descending through 840 feet msl, but incredibly, no comments were made by any crewmember, and no action was taken to stop the descent.

"Hydraulics?" queried the engineer, the next item on the checklist. "Uh, landing lights?"

In very close proximity to the terrain and descending rapidly, the GPWS warned, "sink rate, sink rate!"

"Sink rate," reminded the first officer, "okay."

"Two hundred," called the flight engineer, increasingly aware of the proximity of the ground, and perhaps uncomfortable with the airplane's flightpath. At almost the same time, the first officer said, "Let's make a missed approach."

Two seconds later, the flight engineer reminded the captain that the runway was not in sight, just as the first officer again pleaded, "not in sight! Missed approach!"

"Go around!" urged the engineer.

Finally, almost ten seconds after "minimums" was annunciated by the GPWS and four seconds after the first call for a missed approach, the captain disconnected the autopilot and slowly pushed the throttles forward.

"Flaps?" prompted the first officer, anticipating a flap retraction as required for the climbout maneuver. There was no response from the captain.

"One hundred!" yelled the GPWS. "Fifty...forty...thirty...twenty," all sounded in rapid succession. The captain pulled on the control column and advanced the power for all engines, but the action was too little, too late. Pitch attitude increased through eight degrees and the engines started to "spool up," or increase in speed. But still three-and-a-half miles from the airport, KAL 801 slammed into the side of Nimitz Hill, only yards from the Nimitz VOR.

Clipping trees at the 675-foot msl level, the number one engine (outboard engine on the left wing) impacted the ground first. A large, above-ground pipeline was severed by the left landing gear, and pieces from the left wing were ripped off as the giant airliner hurtled across the face of the hill. The inboard engine on the left wing tore through an unused ordnance bunker, but remained with the wing. Two seconds later, the underside of the fuselage was torn away, scattering cargo and baggage along the debris path.

As the aircraft broke apart, the forward-most nose section, including the cockpit, detached from the rest of the airframe and tumbled down the far side of the hill. As the left wing separated, the right wing and center section rotated 180 degrees from the original heading, coming to rest on top of the left wing. The fuel tanks exploded upon impact and the resultant fire burned intensely for many hours. The forward part of the fuselage came to rest close to the wings, with one section engulfed in the inferno. The tail section came to rest upright, although tilted, and was not completely consumed by the fire (see Fig. 9-4). The total wreckage distribution path, from the initial tree strike to the last documented piece of the aircraft, part of the cockpit, was only 2,100 feet long.

A hunter and his friend were stalking small game on U.S. Navy property surrounding the Nimitz VOR, when the lights of a very low aircraft startled them. They watched in total disbelief as the giant airliner came directly toward them, passing just twenty feet over

9-4 *View of KAL 801 wreckage looking from Nimitz Hill toward airport. Arrow indicates Runway 6L; cockpit section is down the hill to the left.* . . Courtesy NTSB

their heads. The roar was deafening and the air pressure knocked them off their feet. Awestruck, they saw KAL 801 strike the side of the hill just a few hundred feet away, explode into a fireball, and disintegrate. Panicking and believing there was nothing they could do to help, the two fled the area without assisting the victims or notifying authorities.

Of the 254 occupants on the B747, 225 died at the scene, either by impact or in the resulting fire. Twenty-five passengers and four flight attendants survived, some flung from the wreck still strapped into their seats, others forced to claw their way though twisted, burning debris to safety. Overhead bins collapsed onto those seats that remained inside the airplane and piles of carry-on baggage obstructed potential escape routes. It began to rain heavily, soaking the survivors and making movement out of the immediate area impossible. Several passengers would die in the time it took for rescuers to reach the crash site, and three more would succumb within the following month.

The rescue

The last radio transmission received from KAL 801 was at 1:41 in the morning. After efforts to reestablish contact were unsuccessful, the tower controller alerted PGUM ramp control of a possible downed aircraft. They in turn began emergency notifications at 2:02 A.M., about twenty minutes after the crash. Five minutes later, the Guam Fire Department's Engine Company No. 7, located just three miles from Nimitz Hill, was ordered to the scene. That vehicle was equipped with air brakes, however, and by procedure all the brake lines had been drained earlier that night to prevent excessive buildup of condensation. An additional twelve-minute delay was encountered to allow the air brake system to charge and the engine to warm up.

Finally underway, the fire truck reached the security gate at the base of the hill at 2:34 A.M. Only one access road led up to the crash site, and it was very narrow and lined by drainage ditches. Although Engine No. 7 was the first vehicle to arrive, it was immediately joined by units from the local police, U.S. Navy security, Federal fire departments, ambulances, other Guam Fire Department trucks, Anderson Air Force Base vehicles, the on-scene commander, and even the Governor of Guam. Police were quickly pressed into service to direct and coordinate all the traffic at the gate, which was clearly hampering efforts to reach the victims.

The terrain was treacherous and eight-foot-high sword grass made foot travel through the underbrush almost impossible; rescuers quickly realized that the road provided the only means to the site. Engine No. 7 encountered a large section of damaged pipeline blocking the road only one hundred yards from the burning airplane, and immediately got stuck in the mud trying to maneuver around it. Unsuccessful in trying to move the pipe by hand, a winch was brought in, finally clearing the way.

Rescuers rushed down the embankment in their search for survivors, carrying flashlights, ropes, and small trauma kits to the burning hulk. Small groups of survivors were found huddled together; screams from those less fortunate filled the night air. Reaching the site at about 3:00 A.M., medical personnel set up two primary triage areas to care for the injured. After receiving initial treatment, the victims were hand-carried up the steep incline in darkness and driving rain to waiting ambulances, an extremely difficult process. At one point, fire hoses were laid down the slope and were used to drag stretchers back up the hill. Only one ambulance at a time could traverse the narrow access road, however, severely restricting evacuation efforts. It wasn't until much later that military helicopters were used to airlift some of the injured to local hospitals. The last survivor was found at 4:30 in the morning, and the last two injured passengers were extricated from the twisted nose section of the airplane at 7:00 A.M., almost five-and-a-half hours after the accident. No attempts were ever made to suppress the fires at the site, which eventually burned themselves out late the next day.

Removal and identification of human remains was an equally arduous task. Firefighters and Navy Seabees provided the heavy equipment to access the interior of the wreckage, by cutting, chopping, and lifting, while a second group of workers removed light debris. Photographers then documented the scene and medical personnel tagged the remains, plotting them on a master map for reference. Finally, the last group of technicians removed the victims from the site, up the almost impassable terrain to vehicles waiting at the top of the hill.

The investigation and findings

An investigative team led by the National Transportation Safety Board (NTSB) and comprised of representatives from the Federal Aviation Administration (FAA), Boeing, Korean Air Co., Ltd., the National Air Traffic Controllers Association (NATCA), Barton ATC International, and the Government of Guam converged on the scene. Initial indications were that this might be yet another Controlled Flight Into Terrain (CFIT) accident (for additional discussion, see Chapters 2, 3, 13, and 14). Attention immediately focused on weather, flight crew actions and decision-making capabilities, Air Traffic Control (ATC) facilities and handling, design of the instrument approach, KAL training and oversight, and the effectiveness of the response by rescue personnel.

The weather

Rainshowers associated with a weak low-pressure trough existed in the Agana area at the time of the accident, and as they moved over higher terrain, they increased in intensity. Overall, flight visibility was very good, but a broken layer of clouds below 5,000 feet would have restricted the pilot's view of terrain while maneuvering for the final approach. A particularly heavy storm cell was centered only four nautical miles southwest of the airport, directly in the approach corridor to Runway 6L.

Investigators theorized that about eight miles from the runway, flight 801 likely entered clouds and light precipitation. Most likely, the rain increased to a very heavy level at about the outer marker. Significantly, investigators believe that after initially sighting the airport, the crew lost sight of the runway in subsequent heavy rain. Even after exiting that squall, they were unable to see airport because of another rain shower located on the approach course to the northeast of Nimitz Hill.

Flightcrew actions

Thirty minutes before their planned arrival, the captain conducted the required approach briefing. He pointed out that the glideslope portion of the ILS Runway 6L was out of service, but since the visibility was good (six miles), "we are in the visual approach." He further stated that if a missed approach was necessary, they would "stay visual" because the weather conditions were VFR (visual flight rules).

The Board acknowledged that the captain noted the proper glideslope out MDA, but he failed to review the mandatory intermediate altitudes or unusual DME (distance measuring equipment) features of the nonprecision approach. Expecting good weather conditions, an early sighting of Guam only reinforced his belief that a visual approach could be flown. The NTSB believed that this expectation led to inadequate crew preparation for a complex instrument procedure. Additionally, the lack of an effective briefing "set a cockpit tone that made it difficult for other crewmembers to participate in or questions his actions," the NTSB stated. The Safety Board was also concerned that a KAL video training aid produced to assist pilots in flying into Guam emphasized only the visual approach aspects of the area and did not mention any of the unique procedural or terrain features that made operations into the airport challenging.

Confusion about the glideslope

Dispatch release NOTAMS (notices to airmen), the Guam airport ATIS (Automatic Terminal Information Service) information, and ATC controller advisories all alerted the crew to the fact that the glideslope was unusable. Nevertheless, the CVR recorded several cockpit discussions during the descent and approach indicating confusion among the flight crew as to whether the system was providing accurate vertical guidance information.

Investigators found that spurious radio signals may have caused some erratic movement of the glideslope pointers in the captain's and first officer's navigational instruments, suggesting normal reception of the ILS. But distinctive "OFF" flags would have biased into view, covering the pointers and verifying the fact that the signal was not usable. The NTSB suggested that the captain's preoccupation with the glideslope signal may have caused him to overlook important cues that proper approach procedures were not being followed.

The first officer's incorrect "approaching fourteen hundred" call went unnoticed, and the captain directed him to reset the desired altitude to 1,440 feet. Since the aircraft was still descending through 2,300 feet, that action prevented the autopilot from leveling the airplane at the required 2,000-foot altitude. Shortly thereafter, the flight was cleared to land and investigators felt that the crew probably then set the MDA of 560 feet in the altitude window. Again, since the autopilot had not yet captured the intermediate level-off altitude of 1,440 feet, the descent continued.

The Safety Board concluded that the captain's confusion regarding glideslope indications, his neglecting to cross-check raw data glideslope information, and his failure to detect the first officer's incorrect altitude call "resulted in the airplane descending below the minimum altitudes of the step-down fixes prescribed by the approach."

The approach procedure

The ILS to Runway 6L is an unusual approach for several reasons (see Fig. 9-2). The Nimitz VOR/DME, located 3.3 miles from the end of the runway, is used as a "step-down fix" on the localizer approach. In order to utilize this fix, the crew was required to tune the localizer on one navigational radio and the Nimitz VOR on the remaining navigation radio. Thus, typical redundancy of having two nav radios set to the same navigational aid was lost.

As the nonprecision approach is flown, the DME distance should count down to station passage, and then count back up again to the missed approach point, 2.8 DME. This type of approach is typical of a VOR approach, but is rare among localizer approaches and is not encountered in any of the KAL instrument training scenarios. Investigators believed that the off-airport location of the VOR may have misled the crew into thinking that they were on the final step-down approach segment (going away from the VOR) instead of the intermediate segment (proceeding toward the VOR).

Failure to respond to the GPWS

Because KAL 801 was in the landing configuration (landing gear down and flaps extended), the "pull up" and "terrain, terrain" mode warnings of the Ground Proximity Warning System (GPWS) were inhibited during their final approach. At least four other audible system alerts were sounded by the system, however, with no resultant flight crew action. The "one thousand," "five hundred," "minimums," and "sink rate" warnings should have immediately informed the crew of the increasingly precarious flight path of the aircraft. The "sink rate" annunciation, based on aircraft closure rate with the terrain and not vertical speed, should have particularly concerned the crew. However, it is likely that the first officer considered only the airplane's vertical speed when he advised the captain, "sink rate okay."

The Board found that, based on CVR and FDR (flight data recorder) data, the captain "failed to take timely and appropriate action to prevent the airplane from impacting terrain even when confronted with repeated GPWS warnings, and then failed to initiate an aggressive missed approach when he announced his intention to go around."

The decision-making process

As mentioned, the NTSB concluded that the captain likely believed until the last moments of flight that he would eventually encounter visual conditions. Concentrating on obtaining the necessary visual cues to complete the approach, he ignored or missed other signs (increasing precipitation, aircraft position, altitude in relation to the localizer approach, GPWS warnings, etc.) that would have mandated a missed approach. The Board surmised that stress caused by the ambiguous nature of the weather combined with confusion surrounding the usability of the glideslope might have also contributed to errors in the captain's decision-making process.

The investigation concluded that the other two crewmembers failed to take a proactive role in questioning decisions made by the captain, who was the flying pilot. The Board determined that this could partly be due to ineffective Crew Resource Management (CRM) training by KAL and an "advocacy" mentality, or a reluctance to question an authority figure, found in the Korean culture.

Flightcrew fatigue

Cockpit conversation indicated the captain was tired. His "really...sleepy" statement may have indicated a "significant performance degradation," according to the NTSB. Slow reaction times (as evidenced by the GPWS warnings), fixation on one piece of information to the exclusion of others (status of the glideslope), and poor judgment are all indicators of flight crew fatigue. Investigators believed that the captain's decision to continue the approach despite the deteriorating weather may have also been influenced by fatigue. The sooner they arrived, the longer their on-ground rest would be. Therefore, the Safety Board concluded that the captain's fatigue "degraded his performance during the approach and also caused him to feel additional pressure to continue the approach, so he could maximize the available rest time before his planned return flight to Seoul."

Out-of-date approach charts

Five Guam instrument approach charts were discovered in the cockpit wreckage of the 747. For the approach being flown, the Runway 6L ILS, two charts with different dates, minimum-crossing altitudes, navigational fix names, and missed approach procedures were found. Investigators believe that the captain was using the correct, current chart, while the first officer was referencing an out-of-date one. Further, there was no CVR record of the pilots ever comparing the "effective date" information available on the charts. The Safety Board found that the differences between the two procedures were significant enough to warrant some confusion, and may have decreased the effectiveness of the first officer in monitoring the captain's adherence to the proper altitudes during the approach.

Flightcrew training

KAL operates two types of 747s. The 747-300 was referred to as the "classic" model since it had the traditional 747 three-person cockpit crew and flight instrument configuration. The 747-400 was distinct from the "classic" because it had a highly automated, "all glass" cockpit and was configured for two cockpit crewmembers. KAL's "classic" 747 initial, upgrade, and transition training included ten simulator sessions and a simulator check ride. The airline's policy was that all simulator scenarios containing instrument approaches (four ILS and one VOR/DME to runways at Kimpo Airport) were to be flown exactly as dictated in the curriculum; no provisions to vary any of the scenarios were available to the instructors, nor was it encouraged. Because all DME training was to on-airport VOR facilities, no crews were ever trained in the "count-down then count-up" procedure.

The Safety Board noted that training at KAL was accomplished strictly by repetition. "Korean Air pilots were not exposed to varied or complex approach procedures, and were not trained to be adaptable to changes in approach procedures." They concluded that this training "encouraged pilots to rely on memory rather than referencing and cross-checking specific procedures..."

Air traffic control

Two minor ATC errors were discovered during the investigation. When cleared for the approach, the CERAP controller did not provide the flight crew with their position relative to the final approach fix, in this case the outer marker, as required (interestingly enough, this ATC procedure was the result of a 1974 air carrier accident). Additionally, the tower controller did not inform KAL 801 that they were "not in sight" when cleared to land. In both cases, the NTSB noted that the flight crew might have questioned their position if the information had been relayed.

MSAW

The Minimum Safe Altitude Warning (MSAW) system is a ground-based radar advisory that allows air traffic control to warn pilots that they are descending, or are predicted to descend, below a prescribed minimum safe terrain-avoidance altitude, in a particular geographic area. Originally installed in the Guam CERAP in 1990, numerous false (sometimes referred to as "nuisance") warnings limited the effectiveness of the system. A new software package was developed for the facility to inhibit all alerts within fifty-four nautical miles of the radar site, and became operational in February of 1995. No MSAW warnings were generated, therefore, for KAL flight 801 during it's premature descent.

Investigators conducted an MSAW simulation using a fully functional MSAW system (noninhibited) and the flight path of KAL 801, with predictable results. These tests showed that both an aural "low-altitude" and visual warning would have been generated to the controller had the system not been inhibited. These signals would have been triggered as the airplane was descending through 1,700 feet msl, or just over a minute before striking the side of Nimitz Hill. Since the flight was under the control of the tower facility by then, the CERAP controller would have had to notify the tower controller for the alert to be issued, but the NTSB believed that sufficient time was available for the warning to have been effective.

The Safety Board was critical of the Federal Aviation Administration (FAA) in their oversight of the Guam MSAW. Although the inhibition was known as early as July 1995, an FAA inspection failed to produce corrective actions. A second inspection early in 1997 reported the inhibition as an "informational" item only, and again no remedial action was taken. The system was restored to fully operational status shortly after the accident, but the Safety Board felt that "it is very likely that this inhibition would have continued if this accident [KAL 801] had not occurred."

KCAB oversight of KAL

The Korean Civil Aviation Bureau (KCAB) is a division of the Ministry of Civil Transport (MOCT) within Korea and is responsible for all oversight of all aspects of the airline's operation. In evaluating the effectiveness of the KCAB's inspections, the NTSB found "little meaningful surveillance of Korean Air's pilot training program, [and the KCAB] was ineffective in it's oversight of the...pilot training program."

KAL accident history

The NTSB noted that in the last twenty years, KAL has experienced a number of accidents related primarily to pilot performance. These have resulted in over 700 deaths, and include:

- KAL 007, August 31, 1983. Shot down by a Soviet fighter off Sakhalin Island; the investigation concluded that the crew likely made navigational errors resulting in entry into restricted airspace.
- KAL 084, December 23, 1983. A ground collision at Anchorage, Alaska, resulted in substantial damage and three serious injuries when the KAL pilot became disoriented, failed to follow accepted taxi procedures, and took off even though unsure of his position on the airport.
- KAL ML-7328, July 27, 1989. Seventy-eight fatalities resulted from a crash during a nonprecision approach. The cause was determined to be "improper flight crew coordination likely influenced by the affects of fatigue."

- KAL Airbus, August 10, 1994. The aircraft was destroyed during a runway overrun accident, although all occupants survived the crash. The cause was an apparent misunderstanding between the pilots as to whether to continue the approach or go around.
- KAL HL-7496, August 5, 1998. The aircraft crashed during landing in heavy rain at Kimpo Airport in Seoul. No fatalities.
- KAL HL-7236, September 30, 1998. Another overrun accident, this one at Ulsan, Korea, injuring three passengers.
- KAL HL-7570, March 15, 1999. The aircraft overran the runway at Pohang Airport, Korea, and impacted an embankment. Twenty-six passengers were injured.
- KAL 6316, April 15, 1999. Six minutes after takeoff, the aircraft crashed into a residential area of Shanghai, China, killing all three crewmembers and four people on the ground. The accident is still under investigation.

On October 9, 1998, after seven of the more recent incidents, the MOCT imposed what it called "most severe" penalties, including a forced reduction in both domestic and international service. KAL also stated that it planned a comprehensive restructuring of its pilot training programs and review of operating practices and procedures. What action has been taken to date by the airline is unknown.

What is known is that the KAL Deputy Director of Flight Operations testified at the Safety Board's public hearing that prior to the flight 801 accident, "most of our management...has been...short-sighted and superficial in nature...We plan to make long-term plans and spare no resources in [attaining] this final objective. Accordingly, we will adjust our management systems and invest all the more heavily into training and program development." He was fired shortly after his testimony.

FAA oversight of KAL

A Principal international Geographic Inspector (PGI) is assigned by the FAA to provide operational oversight of those foreign carriers operating into the United States. At the time of the accident, the NTSB found that the PGI responsible for KAL also had six other international airlines to observe. And while the FAA does inspect those operations that occur within the United States, they do not "inspect, approve or oversee foreign airlines' training programs or any of their manuals, nor does the FAA accomplish line checks or en route inspections on board foreign airlines." Finally, the PGI told the NTSB that there was no avenue for formal communication between the FAA and the KCAB regarding any aspect of oversight of KAL.

Paragraph 9.3.1 of Annex 6 to the International Civil Aviation Organization's conventions mandates the establishment of an appropriate and effective ground and flight training program. Because of KAL's accident history and the newly discovered deficiencies in the training programs, the NTSB was concerned that the airline may not be in compliance with the international standard.

Emergency response

Investigators determined that the CERAP controller had the ultimate responsibility for monitoring the progress of KAL 801. While he claimed to have been providing that service, the Safety Board believed that had he been more vigilant in his duties, he would have noticed the flight disappear from radar and could have issued a more timely alert to rescue personnel.

Difficult terrain and limited access prevented quick evacuation of the injured from the crash site. But the NTSB determined "a twenty-one minute interval to notify emergency responders and the arrival of the first firefighting equipment fifty-two minutes after the accident to be unacceptable...[and] subsequent deployment of equipment and personnel was inefficient and plagued by preventable delays."

The Safety Board was told at their public hearing that several initiatives were underway to improve future crash, fire, and rescue response and coordination. The first step would be to form a committee to develop a Memorandum of Understanding (MOU) for rescue services to be shared between the airport and the U.S. Navy and U.S. Coast Guard. Next, an airport disaster drill was planned for the fall of 1998, and finally, new radios had been purchased for improved interagency communications and coordination.

As of October 1999, however, no MOU had been signed, nor had any disaster drills taken place. The Board approved of the purchase of the new emergency radios but was concerned that "no concrete action has been taken to improve coordination among Guam's emergency response agencies and that no off airport drills have been conducted in the two years since the accident, although the emergency response to this accident clearly suffered from a lack of inter-agency cooperation and poor coordination of all available re-sources." At the "sunshine" hearing held held in November of 1999, NTSB Chairman Jim Hall called the lack of any remedial effort on the part of Guam's emergency responders "extremely disturbing."

Other issues

As previously recommended by the Safety Board (see Chapters 2 and 3), approach chart terrain depiction and constant angle, stabi-lized-descent approaches were both considered as important safety enhancements, particularly in preventing Controlled Flight Into Terrain (CFIT) accidents. There is currently no FAA requirement for terrain to be illustrated on any approach chart, but because pilots must refer to the profile view during the final approach segment, the Board believed that any possible obstructions or significant terrain should be depicted on that view. Additionally, the constant descent (usually about three degrees) during a nonprecision ap-proach enhances safety by reducing the number of pitch and power changes needed to successfully fly the approach. The Board concluded that these types of approaches would "provide pilots with a more uniform visual perspective outside the aircraft...reduce the difference in descent profiles between precision and nonpreci-sion approaches, provide constant distance/altitude reference points throughout the descent, and enhance the ability of all crewmembers to monitor, cross-check, and make corrections to the glidepath angle."

Conclusions and probable cause

The NTSB found the probable cause of this accident to be the captain's failure to properly brief and execute the nonprecision approach and the flight crew's failure to monitor and cross-check the captain's execution of the approach. Contributing to these failures were Korean Air's inad-equate flight crew training, and the captain's fatigue. Also contributing to the accident was the FAA's intentional inhibition of the MSAW system and the failure of the agency to adequately manage the system.

Recommendations

Safety recommendations were issued to the FAA, the KCAB, and to the Governor of the Territory of Guam.

Those to the FAA included:

- Require that U.S. air carrier POIs ensure that air carrier pilots conduct a full briefing for any available instrument approach that backs up a visual approach at conducted at night or when instrument meteorological conditions might be expected

- Consider designating Guam International Airport a "special" airport, requiring special pilot qualifications.

- Remind pilots, through the *Aeronautical Information Manual*, of the possibility of spurious signals when tuned to navigational aids that are inoperative, and that any navigational indications received must be disregarded.

- Research the benefits of the "monitored" approach, and mandate its use if appropriate.

- Inform all ATC controllers of the circumstances surrounding this accident and remind them of the importance of following established ATC procedures.

- Require the use of vertical flightpath guidance during non-precision approaches on aircraft equipped with systems capable of providing that information, and require, within ten years, that all non-precision approaches approved for air carrier use incorporate constant angle of descent with vertical guidance from on-board navigation systems.

- Issue guidance to air carriers to ensure that pilots periodically perform non-precision approaches during line operations in daytime visual conditions, assuming no additional risk is incurred.

- Evaluate the benefits of terrain depiction and other obstacles on the profile view of approach charts and require such depiction if warranted.

- Provide user groups with draft plan and profile views of new approach procedures to assist them in evaluating those procedures.

- Consider the accident and incident history of a foreign air carrier when evaluating the adequacy of that airline's oversight.

- Require, within two years, that all turbine-powered airplanes with six or more passenger seats not currently equipped with GPWS have an operating enhanced GPWS or terrain awareness and warning system.

To the KCAB:
- Require Korean Air to revise its video training for Guam to emphasize that instrument approaches should also be expected, describe the complexity of such approaches and any significant terrain along the approach path or near the airport.

To the Governor of the Territory of Guam:
- Form a task force, within ninety days, made up of all the emergency response agencies on Guam, to develop emergency notification and response procedures that are timely and effective and to conduct periodic and regularly scheduled interagency disaster drills.

Epilogue

For those that survived the accident, waiting for rescue must have been unbearable. More troubling, though, is the fact that several potential survivors couldn't hang on—lives were lost due to difficulties encountered by emergency personnel. "Unacceptable" was the Board's appropriate conclusion. What is most disturbing, though, is the fact that very little has been done to date to solve this problem.

In their final report, the NTSB reminds the reader of the advances that have been made in efforts to prevent CFIT accidents: Enhanced Ground Proximity Warning Systems, studies done by the Flight Safety Foundation and Airbus, the depiction of hazardous terrain on approach charts, improved nonprecision approach procedure design, crew resource management (CRM), and flightcrew operational procedures to name just a few. The Safety Board has previously issued many, many recommendations in all of these areas, some of which are still pending.

This investigation produced recommendations similar to prior ones, but in this case the emphasis was placed on human performance. The Board recognized that vitally important cues to the progress of the approach went, incredibly, completely unrecognized by the crew until it was too late. Effective pilot training, operationally flexible and culturally focused CRM techniques, and CFIT awareness education will be the hallmarks of future efforts to prevent controlled flight into terrain accidents.

References and additional reading

Buelt, Stanley J. 1998. Mishap investigation of Korean Air flight 801—a team effort. *Flying Safety*. April, 10.

McKenna, James T. 1997. Korean Air rescue flawed. *Aviation Week & Space Technology*. 25 August, 39.

——. 1998. Guam probe cites pilot, FAA errors. *Aviation Week & Space Technology*, 30 March, 59.

National Transportation Safety Board. 1999. *Aircraft Accident Report: Controlled Flight Into Terrain. Korean Air Flight 801, Boeing 747-300, HL-7468, Nimitz Hill, Agana, Guam, Mariana Islands, August 6, 1997*. Washington, D.C.: NTSB.

National Transportation Safety Board. 1999. Public Docket. *Korean Air Flight 801, Boeing 747-300, HL-7468, Nimitz Hill, Agana Guam, Mariana Islands, August 6, 1997. DCA97MA058*. Washington, D.C.: NTSB.

Part Two

Regional Airline Accidents

10

Along for the ride

The final descent of ASA flight 529

Operator: Atlantic Southeast Airlines

Aircraft Type: Embraer EMB-120RT, Brasilia

Location: Carrollton, Georgia

Date: August 21, 1995

The classic definition of an aviation accident, "an occurrence associated with the operation of an aircraft...in which any person suffers death or serious injury or in which the aircraft receives substantial damage," only describes "what" happened. To understand the "how" and "why," it is important to realize that any accident is just the result of a very long chain of events. Breaking any link in the chain could likely prevent the accident.

The flightcrew is usually the last line of defense in accident prevention. But sometimes a series of design, manufacturing, repair, and inspection errors create a situation where the pilot, as much as any passenger on board, is just "along for the ride."

Flight history and background

Atlantic Southeast Airlines (ASA) is a regional airline, typically flying smaller-capacity turbopropeller aircraft from major "hub" cities into smaller "feeder" community airports. High aircraft utilization is one key to profitability, so ground time is kept to a minimum.

ASA flight 529, an Embraer EMB-120 "Brasilia" twin-engine twenty-nine-passenger airliner (see Fig. 10-1), was built in Brazil and delivered new to the airline early in 1989. The Macon, Georgia, based flightcrew operated this aircraft, N256AS, into Atlanta, Georgia, (ATL), earlier in the day and set about their routine duties to quickly prepare for their flight to Gulfport, Mississippi (GPT).

The captain, forty-five years old, had been with the airline for seven years. His total flying time was almost 10,000 hours, with over 7,000 of that in the EMB-120. He remained in the cockpit during the ATL ground stop to receive the air traffic control (ATC) clearance and to complete cockpit preflight checks prior to departure.

The first officer was twenty-eight and had been hired by ASA only four months previously. All of his flying time at the airline, 363 hours, was in the Brasilia. He deplaned in ATL but remained in the immediate ramp area to attend to normal flight planning and passenger boarding duties.

The one-hour, twenty-six-minute flight would be flown at 24,000 feet (FL240). Taxiing away from the gate shortly after noon, flight 592 took off at 12:23 P.M. (EST), with twenty-six passengers, two pilots, and a flight attendant aboard.

Initially, rain and fog limited forward visibility, but the airplane climbed rapidly through the low overcast. A turn towards the southwest was made, the aircraft was configured for a normal climb to cruise altitude, and all checklists were completed. Light turbulence was encountered, but a quick check of the weather radar confirmed that the ride would eventually smooth out.

Twenty minutes after takeoff, while climbing through 18,100 feet at 160 knots indicated airspeed, the cockpit voice recorder (CVR) area microphone picked up the sound of several loud thuds. To the pilots it sounded more like "someone had hit a trash can with a baseball bat." The left engine stopped producing power and the aircraft quickly pitched down and rolled to the left.

"Autopilot, engine control, oil," warned the automated, mechanized voice. "Autopilot, engine control, oil, autopilot..." was repeated continuously.

"We got a left engine out! Left power lever, flight idle." The captain, wrestling with the controls to stabilize their flightpath, started the

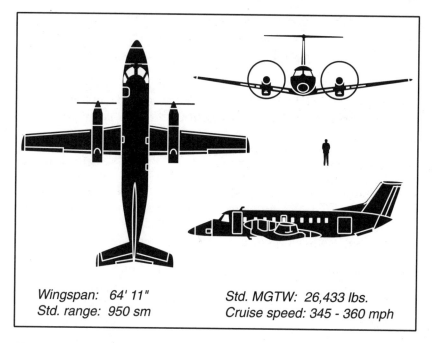

Wingspan: 64' 11"
Std. range: 950 sm

Std. MGTW: 26,433 lbs.
Cruise speed: 345 - 360 mph

10-1 *Embraer EMB-120.*

emergency checklist. The nose of the airplane dropped to about nine degrees down, and the rate of descent increased to over 5,000 feet per minute.

For the next thirty seconds, the plane shook violently. "Left condition lever...feather." The captain was trying to "feather" the propeller blades, streamlining them with the airflow to decrease drag. "Yeah, we're feathered. Left condition lever, fuel shutoff," he continued.

But twenty-five seconds later the captain was still having great difficulty controlling the airplane. "I need some help here!" Almost before the first officer could respond, the captain again cried out "I need some help on this!" He couldn't understand why the airplane continued to descend so rapidly and was difficult to control.

"You said it's feathered?" Perhaps the propeller was still windmilling. "What the hell's going on with this thing?" And a few seconds later, "I can't hold this thing...help me hold it!"

The passengers also felt the airplane shudder, heard the loud bang, and noticed the unusual descent. Those on the left side could see that the engine was askew in its mounts and the front part of the cowling was mangled. The propeller had been moved outward and three of the four propeller blades were wedged against the front of the wing.

"Atlanta center, ASE five twenty nine, declaring an emergency. We've had an engine failure. We're out of fourteen two at this time," radioed the first officer to ATC.

"ASE five twenty nine, roger, left turn direct Atlanta," was the response.

After pulling the power back slightly on the right engine, aircraft attitude was somewhat easier to maintain. "Alright, it's, it's getting more controllable here...the engine...let's watch our speed," cautioned the captain. "Alright, we're trimmed completely here."

The first officer notified the flight attendant that they had declared an emergency and would be returning to Atlanta. ATC asked flight 529 if they could level off, to which the first officer replied "OK, we're going to need to keep descending. We need an airport quick and uh, roll the trucks and everything for us."

"West Georgia," replied the controller, "The regional airport is at your...ten o'clock position and about eight miles."

The captain called for the engine failure checklist as ATC issued headings toward the airport. The first officer started the onboard auxiliary power unit (APU), completed the checklist, and responded several times to new ATC heading assignments and altitude requests. Things were happening fast. Information about the runway and approach were passed along to the crew, and the flight was given a frequency change. Passing through 4,500 feet (msl), the aircraft was still descending at over 1,800 feet per minute.

Looking out his side window, the captain realized why it was impossible to maintain altitude. "Engine's exploded! It's just hanging out there," he said to no one in particular. Because they had descended below cloud level, he could see the ground and asked for vectors for a visual approach to the runway. But they were dangerously low, at only about 800 feet above the ground, and coming down fast.

Still four miles from the airport, the ground proximity warning system (GPWS) started its characteristic warning, followed by the artificial voice reminding "too low, gear!" Ten seconds later, the captain pleaded "Help me, help me hold it, help me hold, help me hold it!" They would be his last words.

The flight attendant carefully prepared the cabin for a possible crash landing and had just finished briefing each passenger on the proper brace position for the emergency landing when she glanced out a window. Assuming that the pilots would inform her when they were close to landing, she was alarmed to see the tops of trees rushing by and realized that a crash was imminent. Quickly strapping herself into the crew seat, she continued to shout instructions to the passengers.

"Stay down...Brace!" she yelled.

A loud stall warning horn blared through the cockpit speakers, followed quickly by activation of the stick shaker. Both pilots strained against the control column, as though sheer will could keep the aircraft flying.

The left wing dipped as the airplane flew through a stand of pine trees, slammed into the ground, and slid about five hundred feet through an open field. The front left side of the aircraft was crushed and as the left wing disintegrated, the left engine was hurled forward of the airplane while smaller pieces were scattered all along the flight path. Most seats were dislodged from the floor and the overhead bins opened, throwing luggage throughout the cabin. The right wing remained attached to the fuselage, which separated into three sections: the cockpit and main cabin areas, both of which remained upright; and the tail section, which came to rest on its right side, directly on top of the aft emergency exit. Only nine-and-a-half minutes had passed since the first ominous sounds were heard in the cockpit.

The rescue

Immediately after the airplane stopped rolling and sliding, passengers started to scramble out of the wreckage. The large breaks in the fuselage, one in the front by the main door and one at the rear of the wing, allowed easy exit for those who could extricate themselves. Sparks crackled from torn electrical cables, and within a minute flames and thick black smoke quickly engulfed the entire cabin area. Those that could not evacuate quickly had to run through the inferno to escape, catching their clothing on fire as they fled. The flight attendant received second-degree burns on her ankles and legs, a broken wrist and collarbone, but continued to move passengers away from the aircraft until the heat became too intense. As other passengers ran from the rear section of the aircraft, she directed them to safety and struggled to extinguish their blazing clothing and skin.

The cockpit had taken the brunt of the impact forces, and the left side and floor areas were crushed. The captain had received severe traumatic injuries to his face and head and would never regain consciousness. Wedged in the wreckage, the first officer found that both the cockpit door and his sliding emergency exit window were completely jammed and would not budge, thus preventing his escape. Reaching behind his seat, he removed a small, wooden-handled crash ax and attempted to break out the side window. After chopping a four-inch hole in the Plexiglas, it was obvious that he would be unable to remove any more of the window. He passed the ax through to a passenger standing outside, but further efforts to enlarge the opening or break the window were unsuccessful (see Fig. 10-2). Making matters worse, the small ax handle separated from the head. The passenger was still using the wooden handle to try to wedge the window open when firefighters arrived a few minutes later. Unable to break through even with full-size axes, rescue personnel rapidly suppressed the fire, cut their way through the cockpit door, smashed down the seat back, and extricated the first officer through the flames. He survived, but serious burns covered fifty percent of his body.

Three passengers and the captain were not as lucky. They perished either in the wreckage or on the way to the hospital. Another passenger died the next day, one twenty-four days later, two more four weeks later, and the last one died four months after the crash. In addition to the nine fatalities, twelve occupants sustained serious injuries, mostly burns.

The investigation and findings

The National Transportation Safety Board (NTSB) convened their investigation at the site the day after the accident. In attendance were investigators from the Federal Aviation Administration (FAA),

10-2 *Cockpit wreckage of ASA 529.* . . Courtsey Pierre Huggins, ALPA

10-3 *Overhead view of ASA 529 wreckage.* . . Courtsey NTSB

Embraer Aircraft Corporation, United Technologies, the parent company of both the engine manufacturer, Pratt and Whitney, and the propeller manufacturer, Hamilton Standard, the Air Line Pilots Association (ALPA), and others. Passenger interviews quickly focused investigative attention on the left propeller assembly, and after surveying the wreckage, only three full left-hand propeller blades and one stub could be accounted for. After determining that the missing blade separated in flight, a computer-generated trajectory analysis indicated the most likely area to find the missing blade and a search was commenced. About three weeks later, a farmer fourteen miles to the southwest of the accident site discovered the remainder of the blade in high grass, only yards outside the original search area.

The 14RF-9 propeller design

The Hamilton Standard 14RF-9 propeller as fitted on the accident aircraft had a solid, forged aluminum alloy spar as the main structural member, with composite material and foam used to create the airfoil shape. A conical hole, called a taper-bore, was drilled into the base of the spar for weight reduction and balancing applications. On early model propellers, the taper-bores were shotpeened, a process in which the surface of the bore is blasted with glass beads or steel shot to increase resistance to cracking. After the first 431 were manufactured, with FAA review and approval, Hamilton Standard deemed the procedure unnecessary and discontinued its use. To balance the blade, a variable amount of lead wool was usually inserted into the taper-bore and was retained with a bleached cork (see Fig. 10-4).

Original propeller certification demonstrated compliance with all of the Federal Aviation Requirements (FARs) pertaining to vibration and resonant frequency considerations, and as outlined in Advisory Circular (AC) 20-66, "Vibration Evaluation of Aircraft Propellers." A resonant frequency is a naturally occurring frequency at which the blade will vibrate when external energy is applied. To prevent overstressing a propeller, it is imperative that only a minimal amount of time is spent in the RPM (revolutions per minute) range that corresponds to a resonant frequency. For reduced cabin noise during ground operations, the EMB-120 propeller rotates at a relatively low idle speed of fifty to sixty percent of maximum propeller RPM (Np). The upper limit was established to only allow operation below the lowest mode of vibration (resonant or natural frequency) of the propeller, although there is variation from one blade to the next. The

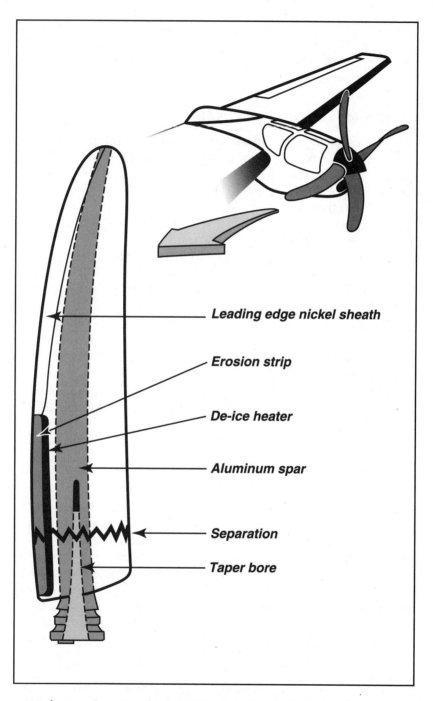

Leading edge nickel sheath

Erosion strip

De-ice heater

Aluminum spar

Separation

Taper bore

10-4 *Hamilton Standard 14RF-9 propeller blade, untwisted planform with section through shank and taperbore.* . . . Source: NTSB Art: J. Watlers

Airplane Flight Manual and a placard in the cockpit cautioned pilots to avoid ground operations above sixty-five percent of maximum Np with the aircraft stationary and a rear quartering wind, to avoid the range of damaging vibrations.

The propeller installed on N256AS

The accident blade was manufactured with an unshotpeened taper-bore in 1989 and had a total of 14,728 operating hours since new. An overhaul was performed by Hamilton Standard about twenty-eight months prior to the accident, at the normal time between overhaul (TBO) of 9,500 hours.

The two blade segments were removed from the scene and examined by the NTSB. They realized early in the investigation that the propeller blade failed as a result of "fatigue cracking [that] appeared to initiate from at least two adjacent locations on the taper-bore surface...and the individual cracks joined to form a single crack that progressed circumferentially around both sides of the taper-bore." Prior to the accident, the crack had progressed through about seventy-five percent of the spar diameter. The Board also noted an area of heavy oxide deposits on the crack faces and on the fracture surfaces near the crack origin. The taper-bore surface also exhibited a series of sanding marks both inboard and outboard of the propeller fracture line, and traces of chlorine were detected on one fracture face.

History of the 14RF-9 propeller

The same condition of taper-bore cracking and oxide deposits was found on Hamilton Standard composite propeller blades from two blade failures in 1994, one in Canada and the other in Brazil. In the first case, the entire propeller and reduction gear box separated in flight and the pilots made a safe landing. In the second situation, the remaining three blades moved toward the "feather" (perpendicular to the airstream for reduced drag) position, also allowing for a successful landing. Additionally, an EMB-120 aircraft in Europe experienced a blade failure while on final approach, but that incident was caused by a fatigue crack on the outside of the aluminum spar about nine inches from the end of the blade.

Because of these earlier incidents, Hamilton Standard began an immediate effort to ultrasonically inspect all applicable in-service propeller blades for cracks. None were found, but abnormal test indications were resulting from "visible mechanical damage" found inside the taper-bore, believed caused by tools used in the insertion and removal of the lead balance wool. To reduce the number of false-positive tests, a procedure was developed by the manufacturer and approved by the FAA that allowed "blending," or very fine sanding inside the bore, followed by application of a special conversion coating and reinspection. If the blade passed one last ultrasonic inspection, it was considered airworthy.

At about the same time, it was discovered that the chlorine used in the bleaching process of the taper-bore cork was a source of corrosion, and in April of 1994, sealant replaced the use of the cork. A series of Hamilton Standard Alert Service Bulletins were also being developed to address the problem of cracks on the inside surface of the taper-bore. In May of 1994, the FAA incorporated these bulletins into a mandatory Airworthiness Directive (AD), 94-09-06, requiring that all blades be "on-wing" ultrasonically inspected for shear wave abnormalities of the taper-bore. Any blade that failed the test was to be returned to Hamilton Standard for repair.

This AD led to the identification of 490 suspect propeller blades, all of which were subsequently sent back to the manufacturer. A few blades exhibited some taper-bore corrosion, and those blades were taken out of service. None of the others, however, had any observable defects, including cracks. All of the returned blades were of the early shotpeened variety, and it was determined that the ultrasonic test was detecting the rough surface of the shotpeened bore, causing a "rejectable" indication. Hamilton Standard engineers decided that the same blending procedures previously used to remove mechanical tool marks would also work to smooth out the shotpeened surface of the taper-bore, thus solving the problem. As Hamilton Standard stated to the Safety Board, the purpose was to "eradicate false ultrasonic positives being caused only by superficial irregularities," and to "remove tool marks or the peaks of shot peen impressions." This procedure was approved and implemented without the knowledge or approval of the FAA or the manufacturer's Designated Engineering Representative (DER).

Another series of bulletins resulted in yet one more AD, 95-05-03, requiring a recurring ultrasonic or borescope inspection of the taper-bore every 1,250 hours. A successful borescope inspection, however, deleted the requirement for any future inspections at 1,250-hour time intervals.

Service history of the accident propeller

The NTSB determined that after its overhaul in 1993, the accident blade was installed on an ASA Brasilia aircraft where it remained in service for over a year. On May 19, 1994, the blade was ultrasonically inspected in compliance with AD 94-09-06, found to have a defect, and returned to Hamilton Standard. According to an internal tracking document, called "a shop traveler form," the blade was reinspected by the manufacturer and again the defect was found. But after the lead wool was removed from the taper-bore, a borescope inspection found no indication of corrosion, pitting, or cracking. The technician noted on the form, "No visible faults found, blend rejected area." Although this taper-bore had not been shootpeened, the blending operation as used for shotpeened bores was completed and signed off on the paperwork. However, the required, "postrepair" ultrasonic inspection was never documented as having been completed. An Airworthiness Approval Tag was affixed to the blade and after rebalancing and other minor work, the blade was returned to ASA. Because no defects were found during the borescope inspection, the blade technically complied with the more recent AD 95-05-03 and no further ultrasonic inspections of the blade would ever need to be done.

Weather

Conditions near the accident site were recorded as a low overcast cloud condition, from 800 feet above ground level (agl) to about 15,000 feet agl, with three miles visibility. Shortly before impact, the crew had the ground in sight and had requested vectors for the visual approach. The local weather report for the West Georgia Regional Airport (CTJ) was not available to the controller handling flight 529, and therefore was not provided to the crew.

Communications

Atlanta Center was the controlling facility at the time of ASA flight 529's propeller separation. There is no dedicated approach control frequency for CTJ, because inbound traffic is handled by Atlanta Approach Control. When the crew requested CTJ runway and airport information for the emergency landing near the end of the flight, Atlanta Center directed them to contact Atlanta Approach. FAA ATC procedures state in part, "If you are in communication with an aircraft in distress, handle the emergency and coordinate and direct the activities of assisting facilities. Transfer this responsibility to another facility only when you feel better handling of the emergency will result." For the ninety seconds that the approach controller was in contact with the flight, he provided a heading to the airport, confirmed the Instrument Landing System (ILS) localizer frequency, and issued a vector for the visual approach.

After declaring the emergency with the Center controller, the flight-crew had asked that emergency equipment be standing by for their landing. That request was never passed along to the Carroll County Fire Department (which provides firefighting and rescue service to CTJ) or to the Atlanta approach controller.

Conclusions and probable cause

The National Transportation Safety Board determined that the cause of the accident was "the in-flight fracture and separation of a propeller blade resulting in distortion of the left engine nacelle, causing excessive drag, loss of wing lift and reduced directional control of the airplane. The fracture was caused by a fatigue crack from multiple corrosion pits that were not discovered by Hamilton Standard because of inadequate and ineffective corporate inspection and repair techniques, training, documentation and communications."

"Contributing to the accident was Hamilton Standard's and the FAA's failure to require recurrent on-wing ultrasonic inspections for the affected propellers."

"Contributing to the severity of the accident was the overcast cloud ceiling at the accident site."

The NTSB found that because Hamilton Standard extended an approved repair to unapproved applications, the accident blade was put back into service with an existing crack. This crack was most likely caused by resonant frequency stresses acting upon corrosion pits inside the taper-bore, and guidance in AC 20-66 regarding separation of propeller vibratory frequencies was inadequate.

The Safety Board also determined that:

- The Atlanta Center controller should have placed the call for emergency equipment when initially requested by the first officer.

- Cockpit-cabin crewmember communication could have been more timely and effective.
- This accident pointed out the need for stricter crash ax design and manufacturing standards.

Party submissions

Embraer, Atlantic Southeast Airlines, the Association of Flight Attendants (AFA), and the Air Line Pilots Association (ALPA) all presented submissions to the Safety Board for inclusion into the public docket. Embraer's and ASA's both sought to correct various technical inconsistencies in the findings of the Powerplant Group. Embraer particularly took issue with certain parts of the NTSB's draft Addendum B, the Powerplant Group Factual Report, that Embraer believed were "technical assumptions" based on the presentation made by Hamilton Standard, not "factual statements."

AFA suggested that guidance from the FAA be made available regarding flight attendant uniforms, primarily to assure that crewmembers can be recognized if no uniform jacket is worn. ALPA urged more intensive oversight of Hamilton Standard by the FAA's Engine and Propeller Directorate. Additionally, party participation in critical component failure investigations was urged, as was research into nonventing oxygen systems, thought to be a possible contributor to the severity of flight 529's fire. Both labor organizations supported changes to AC 120-51B (Crew Resource Management Training) to stress the importance of timely communication between cockpit and cabin crewmembers.

Recommendations

Four days after the accident, the NTSB issued the following safety recommendations:

- Urgent action—Implement an ultrasonic inspection of all applicable Hamilton Standard propeller blades, regardless of previous borescope inspection results. Any blade that had accumulated 1,250 cycles or more since the last inspection was required to be checked before further flight. (A-95-81)
- Conduct a frequency and vibration analysis of EMB-120 propeller blades to confirm that in-service blades retained original certification frequency characteristics. (A-95-82)

- Review the adequacy of maintenance and inspection procedures for all shotpeened taper-bore blades. (A-95-83)

Analysis of the flight data recorder (FDR) indicated a degradation of rudder pedal position data due to an insecure potentiometer (sensor) coupler. So in June of 1996, the Safety Board published two more safety recommendations:

- Review the design of the EMB-120 flight data recorder system, particularly in relation to sensor failures. (A-96-33)

- Require an FDR calibration test every six months until improvements to the sensor's design or installation were implemented. (A-96-34)

The following eight recommendations to the FAA were issued in conjunction with the publication of the aircraft accident final report, sixteen months after the accident.

- Require Hamilton Standard to review the adequacy of tools, training, and procedures for propeller blade repairs, and to assure that those repairs were done properly. (A-96-142)

- Review the possibility of requiring FAR Part 145 Repair Stations to inspect all work done by uncertified mechanics. (A96-143)

- Revise AC 20-66 to include blades that have been in service for a substantial amount of time or have been altered to their maximum limit, and to provide appropriate guidelines for adequate RPM margin between a blade's natural frequencies and normal operational excitation frequencies. (A-96-144)

- Require Hamilton Standard to amend their Component Maintenance Manual inspection procedures to a more appropriate time interval for taper-bore corrosion inspection. (A-96-145)

- Require Hamilton Standard to review and revise its procedures governing internal communication and documentation of engineering decisions, as well as the involvement of the DER and the FAA in that process. (A-96-146)

- Disseminate information to all air traffic controllers, reinforcing the necessity of notifying appropriate crash, fire, and rescue personnel upon a pilot's request for emergency assistance. (A-96-147)

- Update AC 120-51B (Crew Resource Management Training) to include the importance of time-critical communications between cockpit and cabin crewmembers. (A-96-148)
- Evaluate the necessary functions of the on-board crash ax and provide guidance for its design and construction. (A-96-149)

Industry actions

The same day Recommendation A-95-81 was published, the FAA issued a telegraphic Airworthiness Directive (AD), T95-18-51, requiring that any EMB-120 or similar propeller blade that had previously been found to have ultrasonic crack indications be removed and replaced within ten flight hours. The FAA also conducted a "vibration and loads" survey of EMB-120 propellers that substantiated previous certification test results. However, AD 95-25-11 was issued, requiring a cockpit placard reading, "Avoid Np [propeller RPM] above 60% during ground operation." The FAA and the NTSB agreed that it was necessary to limit propeller ground operations to an RPM range that would avoid high vibratory stresses.

The FAA completed the recommended review of Hamilton Standard overhaul and inspection procedures in July of 1996. AD 96-08-02 required repetitive ultrasonic inspection of all EMB-120 blades until restored to original strength and all work was required to be completed by August of 1996.

All Brasilia operators conducted FDR potentiometer calibration tests by November of 1997, but no design defects in the rudder pedal position sensors or attachment mechanisms were noted. It was discovered, though, that a very thin oxide film could build up on the windings of the potentiometer, creating electrical "noise" and unreliable readings. Consequently, Embraer issued a Service Bulletin to provide information on proper potentiometer inspection and repair procedures.

To develop a long-range response to the problem, the FAA met with representatives of Hamilton Standard early in 1997 and came up with an "action plan" to address issues of propeller blade repair and corporate communications deficiencies raised by the NTSB. Hamilton Standard also conducted "Special Supplemental Audits" of all repair facilities. The following have been implemented:

- The blending repair for taper-bores is no longer an acceptable repair procedure.

- Special Audits of all Hamilton Standard repair facilities will be conducted at least once a year.

- A new review process is in place, called the "Safety Parts Program." It is applicable to all training, procedures, and tools for critical propeller component repairs, and features a heightened awareness of the necessary involvement of the DER and the FAA in the design, manufacturing, and repair of these parts.

- AC 20-66, "Vibration Evaluation of Aircraft Propellers," has been revised to include all propeller life and physical limit issues as recommended by the NTSB.

- Hamilton Standard has developed a taper-bore corrosion control program covering inspection techniques, flight hour/calendar time intervals, and repair procedures. This program will become part of the approved Propeller Component Maintenance Manual.

In reviewing the requirements for Part 145 Repair Station employees, the FAA found that the current requirements are adequate. All FAA Principal Maintenance Inspectors (PMIs) were directed to ensure that all stations comply fully with 145.39 (a), regarding qualifications and abilities of uncertificated repairmen.

The FAA went beyond the recommended actions regarding reminding air traffic controllers of the importance of timely notification of crash, fire, and rescue services upon a pilot's request. All ATC facility managers were required to review the regulations and ensure that current emergency notification telephone contacts and procedures are available to each controller. Further, all facility personnel were briefed on the location of that information and the importance of forwarding the status of all emergency aircraft to supervisors and other facilities that might be involved.

On October 30, 1998, after almost three years of study, the FAA issued the rewritten AC120-51C, "Crew Resource Management Training." One section of this publication specifically addresses effective communication of time management information between crewmembers, particularly in emergencies.

Finally, the Society of Automotive Engineers (SAE) S-9 (cabin safety) committee is working to develop an "aerospace standard" for the design and manufacture of all aircraft crash axes. As an interim measure, the FAA has published an AC to assist users in the selection of a suitable ax and as an aid in developing a future standard.

Epilogue

The simple answer, a failed propeller, is only part of the "why" of this crash. Recommendations and subsequent action will prevent similar propeller separations. But in reality, it was the system that failed. The usual checks and balances of design, inspection, repair, certification, and oversight were compromised, and nine people died. Systematically integrating safety into each step of the process could have broken every link in this unfortunate chain of events.

References and additional reading

Federal Aviation Administration. 1998. *NTSB Recommendations to FAA and FAA Responses Report*. <http://nasdac.faa.gov>

Flight Safety Foundation. 1997. In-flight separation of propeller blade results in uncontrolled descent and fatal accident to twin-turboprop commuter aircraft. *Accident Prevention*. February, 1-11.

Garrison, Peter. 1997. Blade failure. *Flying*. May, 106-108.

National Transportation Safety Board. 1996. *Aircraft Accident Report: In-flight loss of propeller blade, forced landing, and collision with terrain. Atlantic Southeast Airlines, Inc., Flight 529. Embraer EMB-120RT, N256AS, Carrollton, Georgia, August 21, 1995. NTSB/AAR-96/06*. Washington, D.C.: NTSB.

National Transportation Safety Board. 1996. Public docket. *Atlantic Southeast Airlines, Inc., Flight 529. Embraer EMB-120RT, N256AS, Carrollton, Georgia, August 21, 1995. DCA95MA054*. Washington, D.C.: NTSB.

Phillips, Edward H. 1995. Blade failure focus of NTSB crash probe. *Aviation Week & Space Technology*. 28 August, 31.

Pomerantz, Gary M. 1998. Nine minutes, twenty seconds: A series. *The Atlanta Journal-Constitution*. November, 15-20.

11

Quandary at Quincy

The runway collision of United Express 5925 and Beechcraft King Air N1127D

Operator: Great Lakes Aviation

Aircraft Types: Beechcraft 1900C and Beechcraft King Air A90

Location: Quincy, Illinois

Date: November 19, 1996

The "see and avoid" concept is a fundamental principle of aviation safety and is a primary responsibility of every pilot. The FAA's *Aeronautical Information Manual* (AIM) states that when meteorological conditions permit, regardless of the type of flight plan or whether or not under the control of an air traffic control radar facility, pilots are responsible to see and avoid other traffic, terrain, or obstacles.

Equally critical to flight safety is effective communication. Broadcasting intentions and concerns in a clear and concise manner is essential in preventing those misunderstandings that can lead to disastrous errors.

Aviation mishaps are rarely the result of one single error or set of circumstances, but rather are the culmination of several components. Like links in a chain, if one of these components is removed, the accident sequence will not occur. A tragic chain of events combined on an otherwise quiet afternoon at Quincy, Illinois, when faulty see-and-avoid techniques combined with several instances of poor communications practices. In aviation, we know mistakes can be unforgiving; in this case the results were deadly.

Flight history and background

Around 3:00 P.M. central standard time (CST), United Express flight 5925, a Beechcraft 1900C (see Fig. 11-1), departed Chicago O'Hare International Airport (ORD) for Burlington, Iowa. The flight was running almost three hours late due to a maintenance problem that arose earlier in the day. The aircraft arrived in Burlington at 4:27 P.M. and departed nine minutes later for Quincy, Illinois, on the crew's eighth and final scheduled leg of the day. It had been a long day for the thirty-year-old captain and twenty-four-year-old first officer, as they reported for duty at the Quincy airport at 4:15 that morning. They likely looked forward to completing this short, seemingly routine, hop down the Mississippi River. Once airborne, the captain remarked, "Look at that sunset...that's gorgeous!"

Indeed, it was a beautiful fall evening, with the Quincy Airport reporting a temperature of thirty-four degrees Fahrenheit and twelve miles visibility. There was a 13,000-foot broken cloud ceiling, with a higher overcast layer at 20,000 feet. The winds were from the northeast (060 degrees) at ten knots.

Using the air traffic control (ATC) call sign of "Lakes Air 251," the flight was operated under Part 135 of the U.S. Federal Aviation Regulations (FARs) by Great Lakes Aviation, a regional airline that shared a cooperative marketing agreement with United Airlines.

The Quincy Municipal Airport has three runways, Runway 18/36, Runway 13/31, and Runway 4/22, as shown in Fig. 11-2. Like hundreds of smaller airports in the United States., the Quincy Airport is an uncontrolled airport, meaning that there is no operating air traffic control tower. Pilots of arriving and departing traffic are supposed to "self-announce" their position and intentions on a designated Common Traffic Advisory Frequency (CTAF) frequency. In order to be most effective, it is imperative that pilots carefully monitor the frequency as well. The AIM states, "It is essential that pilots be alert and look for other traffic and exchange traffic information when approaching or departing an airport without an operating control tower."

At 4:52 P.M., the captain of Lakes Air 251 stated on the CTAF that their flight, "a Beech airliner, just about thirty miles to the north of

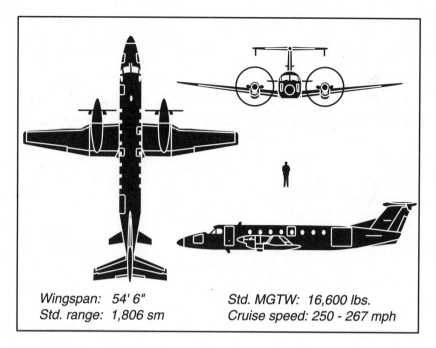

Wingspan: 54' 6"   Std. MGTW: 16,600 lbs.
Std. range: 1,806 sm   Cruise speed: 250 - 267 mph

11-1 *Beechcraft B1900C.*

the field, will be inbound for landing Runway one three at Quincy. Any traffic in the area please advise." The cockpit voice recorder (CVR), which also records radio communications as well as intra-cockpit communications, did not record any replies to the captain's query.

Three minutes later, a female voice identified as one of the occupants of King Air N1127D, announced on the CTAF, "Quincy traffic, King Air one one two seven Delta's taxiing out uh, takeoff on Runway four, Quincy." Shortly thereafter, the pilot of a Piper Cherokee announced that he was taxiing behind the King Air, and stated that he was "back-taxiing" for departure on Runway 4.

In response to hearing these traffic advisories, the captain stated to the first officer, "They're both using [Runway] four" and asked the first officer if he was planning to use Runway 13. The first officer affirmed his plans for landing Runway 13, but added, "if it doesn't look good then we'll just do a downwind for [Runway] four…"

The captain then broadcast on the CTAF that they "were ten miles to the north of the field. We'll be inbound to enter on a left base for

Runway one three at Quincy. Any other traffic please advise." Again, there was no response. The descent checklist was then accomplished, and during that checklist, the crew confirmed that the landing and logo lights were on.

About a minute later, the female occupant of the King Air announced on the CTAF, "Quincy traffic, King Air one one two seven Delta holding short of Runway four. Be, uh, taking the runway for departure and heading, uh, southeast, Quincy."

Upon hearing that advisory the captain of Beach airliner remarked to her first officer, "She's takin' Runway four now?" and the first officer responded, "yeah." The captain then announced on the CTAF, "Quincy area traffic, Lakes Air two fifty one is a Beech airliner currently uh, just about to turn, about a six-mile final for Runway uh, one three, more like a five-mile final for Runway one three at Quincy." The crew then completed the landing checklist.

"And Quincy traffic, Lakes Air two fifty one's on short final for Runway one three um, the aircraft gonna hold in position on Runway four, or you guys gonna takeoff?" announced the captain.

The King Air pilots did not respond, but within seconds, the Beech airliner's CVR recorded a male voice announcing on the CTAF, "Seven six four six Juliet uh, holding uh, for departure on Runway four..." At that moment the Beech airliner's ground proximity warning system (GPWS) blocked a portion of that pilot's transmission by blurting out over the cockpit speakers, "two hundred" to signify that the aircraft was two hundred feet above ground. When the GPWS auto callout was finished, the remainder of that pilot's transmission was recorded on the Beech airliners's CVR as "...on the uh, King Air."

The Beech airliner replied, "OK, we'll, we'll get through your intersection in just a second, sir [unintelligible]. We appreciate that." The crew then completed their final landing checklist.

At 5:00:01 P.M., two seconds after the Beech airliner touched down on Runway 13, the captain exclaimed, "max reverse," no doubt in response to seeing the King Air accelerating on takeoff roll on Runway 4. Leaving 475 feet of continuous skid marks, the Beech airliner collided with the King Air. Both airplanes came to rest with their wings interlocked along the east edge of Runway 13, approximately 110 feet east of where their skid marks converged at the runway intersection.

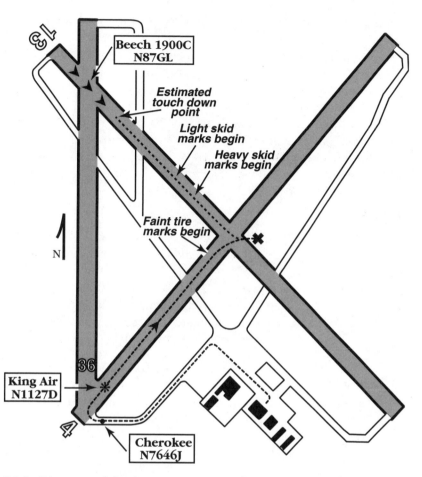

11-2 *Diagram of the Quincy airport and ground tracks of both aircraft...*Source:NTSB

Rescue attempt

Two United Express pilots who had been waiting for flight 5925 to arrive and a pilot employed by the airport's fixed base operator (FBO) were the first to reach the accident scene. They found the King Air totally engulfed in flames, as was the right side of the Beech 1900C. Although they could hear signs of life from the Beech airliner's cabin, they could not see inside because it was filled with dark smoke. The FBO pilot ran to forward left fuselage where he saw the captain's head and arm protruding from her small cockpit side window. Black smoke was pouring from the window, and the captain's blond hair was black from soot. "Get the door open!" shouted the captain.

Responding to that plea, the FBO pilot ran to the forward air stair door and found the door handle in the "6 o'clock" position. Unaware of the exact door opening procedures and unable to see any printed instructions for its operation, the FBO pilot tried desperately to open the door by moving the handle in all directions and pulling on the door. Despite his attempts, however, it would not budge. One of the off-duty United Express pilots then came to assist, and although the handle felt normal as he rotated it, he, too, was unable to open it. He burned his head and ducked away. In desperation, the FBO pilot again tried, but to no avail. The right wing buckled and the fuselage collapsed to the right.

The Quincy Airport had one 500-gallon airport rescue and fire fighting (ARFF) truck. Federal Aviation Regulations stipulated that this equipment be staffed by firefighters fifteen minutes before and after the departure and arrival of any air carrier aircraft with more than thirty passenger seats. However, as the Beech 1900C had only nineteen passenger seats, and there were no larger air carrier aircraft operating at the Quincy Airport at that time, there were no airport firefighters on duty when the accident occurred.

The Quincy Fire Department received a 911 crash notification call at 5:01 P.M. Within one minute, two fire engines with seven firefighters were dispatched to the airport. About twelve minutes later, after traveling a distance of ten miles, the firefighters arrived and began spraying water on the wreckage with 1-inch hose lines. Soon thereafter the county fire department arrived and the combined forces staffed the airport ARFF truck and had the fire under control within about ten minutes.

Meanwhile, trapped in their respective burning airframes, smoke and fumes had fatally overcome the two occupants of the King Air and the twelve occupants of the Beech airliner.

The investigation and findings

The majority of the Beech 1900C's upper fuselage, cockpit, and both wings were destroyed by fire. The nose section was intact, but had been compressed inward approximately ten inches. Blue paint transfer matching the color of the King Air was found along the right side of the nose section. All propeller blades on the left engine remained intact, but each of the right engine's blades were fractured and separated at the blade root. One of these fractured blades was found at the intersection of Runways 13 and 4, and its leading edge had blue paint transfer that matched the color of the King Air's trim paint. The leading edge of another fractured blade exhibited white paint transfer that matched the color of paint on the King Air's fuselage and wing surface.

The King Air's tire scuff marks veered continually to the right on Runway 4 for about 260 feet before converging with those from the Beech 1900C. A fuel spill was evident at the point where the two airplanes' tire marks converged. Although almost all of the King Air was consumed by postcrash fire, a recovered piece of the King Air's radome exhibited chipped paint and marks, and a section of the wing's leading edge showed black paint transfer that matched the color of paint on the Beech 1900C's propellers. No information about the radio's functionality or settings at the time of the accident could be determined from the wreckage. Engine controls of both airplanes were in positions consistent with normal shutdown.

Neither airplane was equipped with a flight data recorder (FDR), nor were they required to have one installed. The King Air did not have a CVR, nor was one required. Therefore, investigators could only speculate what went on in that cockpit in the moments leading up to the collision. The first step in this process was to learn about the King Air's occupants and the details of their intended flight.

King Air's mission and occupants

The King Air had landed at Quincy about 4:25 P.M. and had deplaned several passengers after a sales demonstration flight to potential customers. The pilot was ferrying the aircraft back to its home base, St. Louis Downtown Parks Airport, at Cahokia, Illinois, and accompanied by a passenger who possessed a commercial pilot certificate.

The sixty-three-year-old pilot of N1127D was a retired Trans World Airlines (TWA) pilot, and had served as a pilot in the U.S. Air Force Reserves. In October 1991, TWA transferred this pilot from captain status to flight engineer because of failure to successfully complete a proficiency check and special line check. He left TWA in 1992. From 1993 until the time of the accident, he was a part-time flight instructor at a nearby Air Force aero club, and flew part time as an on-demand air taxi pilot.

11-3 *Overhead view of accident site and positions of victims...*Courtesy AFIP

236

Seven months prior to the accident, the pilot was involved in a wheels-up landing in a Cessna 177RG, while giving flight instruction to a student commercial pilot. The FAA pursued enforcement action against the pilot for this incident, but opted to allow him to complete remedial training in lieu of punitive actions. This remedial training had not been completed at the time of the accident. The FAA inspector who investigated that incident told NTSB investigators that the pilot "expressed an extremely negative attitude toward the FAA's questioning him about the landing. His statements were to the effect that he was a retired U.S. Air Force Colonel with almost 30,000 hours of flying time and that landing gear-up did not mean anything."

The NTSB also interviewed several pilots who had received instruction from the King Air pilot, including the student who was involved in the gear-up landing. The student indicated that in his opinion, one shortcoming of the King Air pilot was that he "seemed to be in a hurry when time was a factor," and that sometimes the instructor tried to rush him. Two of the passengers who had flown with the King Air pilot to and from Tulsa, Oklahoma, just before the accident indicated that the pilot seemed "to be in a hurry" or "anxious to get home." The pilot's wife indicated that she expected the pilot to return home at about 5:45 P.M. that evening. The King Air owner later told investigators that flying time from Quincy to the plane's home base was approximately thirty to forty minutes.

The thirty-four-year-old passenger/pilot of King Air N1127D held a commercial pilot certificate with multiengine, instrument, and seaplane ratings. She also held a certified flight instructor certificate with airplane single and multiengine ratings, and was qualified as an instrument flight instructor and certified ground instructor. She was employed by FlightSafety International as a ground school instructor, and she occasionally flight instructed at the same Air Force aero club as the King Air pilot. Her logbook indicated that she had 1,462 flying hours.

It was reported that the pilot/passenger was interested in building multiengine flight time so that she could qualify as a pilot for a regional airline. She had never flown in a King Air previously, and the King Air pilot offered to take her on this flight to help her build time. One of the passengers on the King Air's previous flights to and from Tulsa noted that both the pilot and pilot/passenger had their hands on the controls when they left Tulsa. Another passenger stated that the two pilots appeared to have "an excellent teacher/student relationship with [the pilot] as the teacher." A third passenger told investigators that "[the pilot] seemed to be telling [the pilot/passenger] how to fly the plane when we were in flight."

The bodies of the King Air occupants were found behind the cockpit seats, so investigators were unable to determine which cockpit seat each pilot occupied during the accident flight.

King Air takeoff distance and accelerate stop time and distance

Using the manufacturer's airplane performance data, the NTSB computed the estimated takeoff distance for the King Air as being approximately 1,500 feet. The pilot of the Piper Cherokee who was behind the King Air prior to its takeoff roll indicated that the King Air taxied onto Runway 4 and waited for about one minute before initiating takeoff roll. He stated that the King Air had pulled forward on Runway 4 just enough so that he could taxi behind it and proceed to Runway 36. From that position on Runway 4, the distance to the intersection of Runways 4 and 13 was approximately 1,900 feet. Investigators also computed accelerate-stop time and distance to pinpoint the exact time that the King Air began its takeoff roll.

Visibility and conspicuity tests

Three days after the accident, investigators performed tests at the accident site to determine line of sight visibility and conspicuity of the Beech 1900C in sunset/dusk conditions. Under similar time and meteorological conditions, two investigators positioned themselves at the approach end of Runway 4 facing the direction of the King Air's takeoff roll. Between 4:37 P.M. and 5:08 P.M., they observed operations of three airplanes that either took off or landed on Runway 13. In each case, they found the landing and strobe lights to be conspicuous from their vantage point on Runway 4, and they noted no visual obstructions that would have affected these aircraft.

Following these tests, two test pilots and an investigator took off in a Beech 1900C and landed on Runway 13, while other investigators observed from the cockpit of a King Air B90 that was positioned on Runway 4 in the approximate takeoff position of the accident airplane. They conducted three tests. The first involved having the Beech 1900C fly overhead, make a downwind entry, and land on Runway 13. During this test, the investigator in the left seat of the King Air noted that the rear window side post obstructed his vision of the Beech 1900 for most of its short final approach and touchdown.

The second test involved positioning the King Air about 100 feet further down the runway from its previous position. For the third test, while the Beech 1900 was about 200 feet above ground level (agl), a test pilot in the King Air's right cockpit seat began a fast taxi toward the accident site. In tests two and three, the Beech 1900 appeared close to the King Air captain's forward window post. No surface obstructions were noted in any of these tests.

The NTSB used manufacturer-provided photographs that represented the views from the cockpits of both airplanes. For each aircraft and each seat position, two photos were superimposed to represent the visual image for each eye and the aircraft windshield visibility angles. These photographs were taken from the "design eye zero reference point" and represented a panorama of what a pilot would see when the head was rotated, but do not account for any movement of the torso, or head movement forward, backward, up, or down.

These photos were then used in conjunction with a computer program that plotted each aircraft's estimated position in relation to the other aircraft's cockpit windows. Diagrams reconstructing the pilot views as represented in those photographs are included as Figs. 11-4 through 11-7. As shown in Fig. 11-4, the Beech 1900 captain's view of the King Air would have been almost completely blocked by the windshield center post and wipers during the airplane's final approach. Further, the view of the King Air would have been totally obscured by the Beech 1900's center post for six seconds, starting at 5:00:43 P.M. From roughly ten seconds before touchdown until the time of the collision, the captain could have had a partially obstructed view of the King Air through the first officer's windshield.

Figure 11-5 shows that the view of the King Air from the Beech 1900's first officer seat would have been at least partially obstructed by the aircraft's right side window post. This obscuration would have lasted from thirty seconds before touchdown until impact.

The field of vision of the King Air pilots would have been somewhat restricted as well. Figure 11-6 illustrates that while the King Air was stationary near the threshold of Runway 4, the left seat occupant's view of the Beech 1900C would have been unobstructed at approximately 72 degrees to the left, as viewed through the King Air's side window. At 5:00:42 P.M., four seconds before the King Air began its takeoff roll, the view of the 1900C from the King Air left seat would have become partially blocked at 68 degrees to the left. At 5:00:54 P.M., while the King Air was accelerating down Runway 4 on its takeoff roll, and just before the 1900C touched down on Runway 13, the view of the 1900C would have become totally blocked, and would have remained blocked until the time of impact.

As shown in Fig. 11-7, the right seat King Air pilot should have been able to see the Beech airliner just before the King Air initiated its takeoff roll. The view of the Beech 1900 would have been blocked shortly after the King Air began takeoff roll, but would have emerged into clear view in the left front windshield at 5:00:39 P.M.— thirty seconds before the collision. The NTSB noted that when the King Air initially was throttled back shortly before the collision, the Beech 1900C would have been about 50 degrees left of the King Air. They concluded that this position would have been fully obstructed from the left seat occupant's position, but in clear view from the right seat occupant's position.

Conclusions

The NTSB concluded that the pilots of United Express flight 5925 were properly certified, trained, and qualified for their duties, as was the pilot of the King Air. The Beech airliner was equipped and maintained in accordance with applicable rules and directives. Investigators could find no records to indicate that the King Air had undergone required transponder and pitot-static inspections, but they concluded that the lack of those inspections did not contribute to the accident. Weather was not a factor in the accident.

Performance of the United Express crew

The NTSB noted that the crew of the Beech 1900 made a "straight in" approach for Runway 13, rather than following the recommended procedures in FAA Advisory Circular AC90-66A, "Operations at Uncontrolled Airports." However, the Safety Board acknowledged that AC90-66A provided strictly advisory material, and that the FAA recognizes that some pilots may conduct straight-in approaches for various operational reasons. For that reason, the FAA states that all pilots

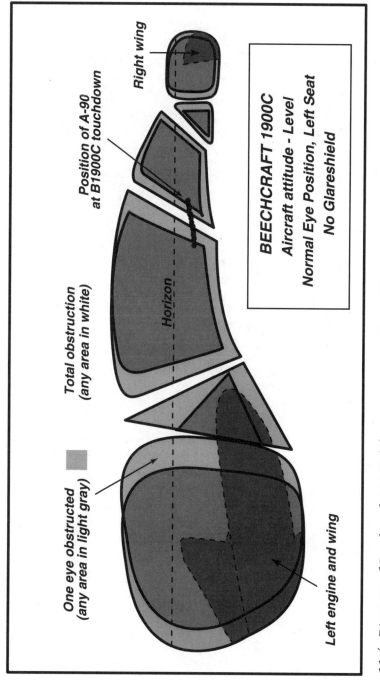

BEECHCRAFT 1900C
Aircraft attitude - Level
Normal Eye Position, Left Seat
No Glareshield

Right wing

Position of A-90
at B1900C touchdown

Horizon

Total obstruction
(any area in white)

One eye obstructed
(any area in light gray)

Left engine and wing

11-4 *Diagram of Beechcraft 1900C left seat occupant's view of King Air.*...Source: NTSB Art: J. Walters

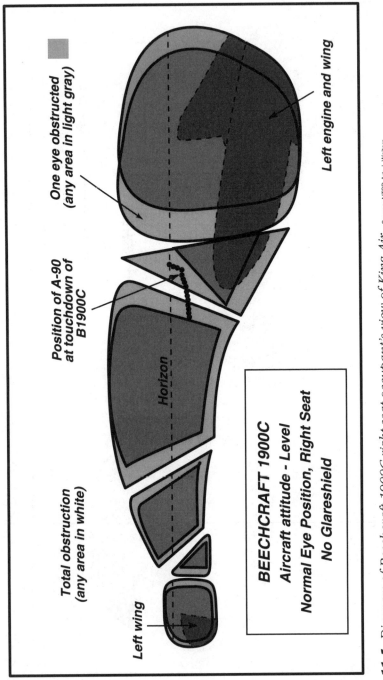

One eye obstructed
(any area in light gray)

Position of A-90
at touchdown of
B1900C

Left engine and wing

Horizon

Total obstruction
(any area in white)

Left wing

BEECHCRAFT 1900C
Aircraft attitude - Level
Normal Eye Position, Right Seat
No Glareshield

11-5 *Diagram of Beechcraft 1900C right seat occupant's view of King Air…*Source: NTSB Art: J. Walters

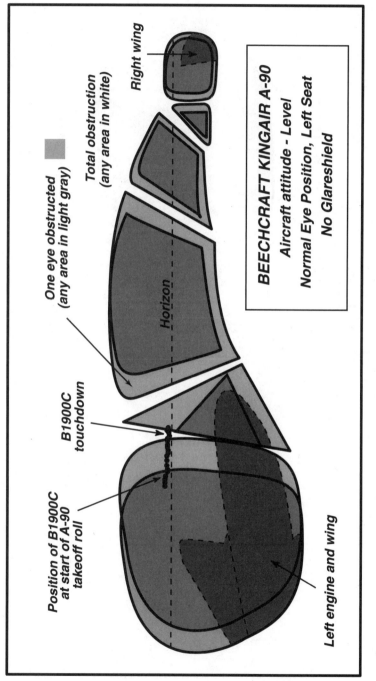

One eye obstructed
(any area in light gray)

Total obstruction
(any area in white)

Right wing

Horizon

B1900C
touchdown

Position of B1900C
at start of A-90
takeoff roll

Left engine and wing

BEECHCRAFT KINGAIR A-90
Aircraft attitude - Level
Normal Eye Position, Left Seat
No Glareshield

11-6 *Diagram of King Air left seat occupant's view of Beechcraft 1900C...*Source: NTSB Art: J. Walters

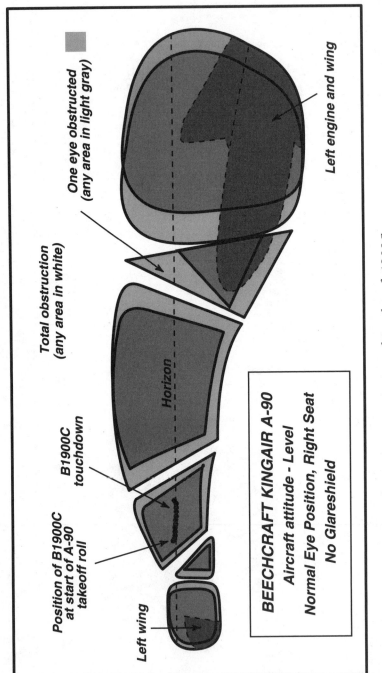

**One eye obstructed
(any area in light gray)**

**Total obstruction
(any area in white)**

Left engine and wing

Horizon

**B1900C
touchdown**

**Position of B1900C
at start of A-90
takeoff roll**

BEECHCRAFT KINGAIR A-90
Aircraft attitude - Level
Normal Eye Position, Right Seat
No Glareshield

Left wing

11-7 *Diagram of King Air right seat occupant's view of Beechcraft 1900C...*Source: NTSB Art: J. Walters

operational reasons. For that reason, the FAA states that all pilots should be alert for other aircraft while operating in the pattern.

According to the NTSB, the wind speed and direction did not particularly favor one runway over the other. They also noted that the Beech 1900 crew discussed that two other airplanes were planning to take off on Runway 4, and they had an alternative plan to use Runway 4 if necessary. The Safety Board concluded that "given the Beech 1900 flightcrew's frequent radio broadcasts of the airplane's position during the approach, and the lack of any prohibition on straight-in approaches to uncontrolled airports, the flightcrew's decision to fly a straight-in approach to Runway 13 was not inappropriate."

In spite of the FARs stipulating that landing aircraft have right of way over other aircraft operating on the airport, the NTSB found that "the captain took the precautions of asking whether the airplane on the runway was going to hold or take off." They found that the Cherokee pilot's radio transmission ("seven six four six Juliet, uh, holding for departure on Runway four, [unintelligible statement due to GPWS callout on the Beech 1900], on the uh, King Air") immediately following the captain's inquiry appeared to be in response to her question. The NTSB said that there were subtle cues that could have alerted the Beech 1900 crew that this transmission did not come from the King Air. These cues included the use of a different "call sign" (the "N" registration of the aircraft), and the gender of the speaker being a male instead of a female, as the King Air's previous transmissions had been from a female. The NTSB concluded that the 1900 crew did not focus on these cues because they were likely preoccupied with landing the airplane, and because this radio transmission stated "King Air" and not "Cherokee." Additionally they noted that the pilots had no reason to expect a response from any other aircraft other than the King Air. "Having received what they believed was an assurance from the airplane on the runway that it was going to hold, the pilots may have become less concerned about continuing to watch the King Air during landing," concluded the Safety Board.

In their final analysis, the NTSB determined that the Beech 1900 crew made appropriate efforts to coordinate their approach and landing through radio communications and visual monitoring. "[H]owever, they mistook the Cherokee pilot's transmission (that he was holding for departure on Runway 4) as a response from the King Air to their request for the King Air's intentions, and therefore mistakenly believed that the King Air was not planning to take off until after flight 5925 had cleared the runway."

Performance of the King Air pilot and pilot/passenger

The NTSB found several indications that the pilot may have been providing flight instruction to the pilot/passenger on the accident flight. These indications included the fact that the pilot/passenger was known to be trying to build multiengine flight time. Additionally, passengers on the previous flights to and from Tulsa indicated that the pilot was providing flight instruction to her on those flights. The NTSB also pointed to the King Air pilot's action of taxiing onto Runway 4 and then holding in position for approximately one minute before beginning takeoff roll. The Board said this action further suggested that the pilot may have been preoccupied with instructional activities.

The NTSB noted that when the King Air pilot initiated takeoff roll, it was done without making an announcement on the CTAF. Given the one-minute delay on the runway, this announcement "would have been prudent, and would have been consistent with common and expected practices at uncontrolled airports," stated investigators. "The Safety Board concludes that the failure of the King Air pilot to announce over the CTAF his intentions to take off created a potential for collision between the two airplanes."

The King Air radios were destroyed by post-crash fire, so investigators could not determine their settings. Investigators were able to conclude that the occupants of the King Air either did not properly set the radios, either in volume and/or to the proper frequency, or that they were preoccupied, distracted, or inattentive.

The NTSB noted that witnesses could see the Beech 1900 while on its extended final approach. However, the NTSB's visibility studies demonstrated that the view of the Beech 1900 would have been partially obstructed by the King Air's cockpit side posts during much of the Beech 1900's final approach and landing, and during the King Air's takeoff roll. In spite of these obstructions, the NTSB pointed out that "those obstructions could have been easily been overcome if the King Air's occupants had moved their heads and bodies while scanning." The NTSB stated that if these pilots would have properly scanned "at any point during the last four minutes of the [Beech 1900's] approach, they would have been able to see the incoming airplane and would not have commenced their takeoff roll when they did." They therefore concluded that the King Air occupants failed to scan for traffic.

Investigators speculated that a combination of preoccupation with providing instruction to the pilot/passenger, careless flying habits, possible fatigue, and rushing could explain why the King Air pilot did not properly scan for traffic. They supported this speculation by noting his past unsatisfactory piloting performance, both at TWA and more recently by being involved in the gear-up landing. They also noted statements of others who had flown with him and indicated that he was sometimes affected by time pressure; he was observed "to be in a hurry" or "anxious to get home" just before they departed on the accident flight.

Performance of the Piper Cherokee pilot

The NTSB noted that the Cherokee pilot only had eighty hours of flight time, and commented that his poor radio technique reflected his inexperience. In his first radio communication that was recorded on the Beech 1900's CVR, the Cherokee pilot was heard stating that he was going to "back taxi to Runway 4." The NTSB stated that he used the term "back taxi" inappropriately, as the term refers to taxiing on a runway opposite the traffic flow.

His poor radio technique later played a role in contributing to the accident's causation, the NTSB determined. At 5:00:16 P.M. when the Beech 1900 captain asked if "aircraft gonna hold in position on Runway 4, or you guys gonna take off?" According to the NTSB, "The Cherokee pilot's response to flight 5925's question (directed to the airplane 'in position on Runway 4') was unnecessary and inappropriate because he was not the first in line for takeoff." The potential for confusion was compounded because the Cherokee pilot did not proceed his "N" number with his airplane model ("Cherokee"), as recommended in the AIM.

The investigation concluded that the Cherokee pilot's transmission, coupled with his failure to correct any misunderstanding that his transmission may have created, misled the United Express crew into believing that they had been communicating with the King Air pilot, and that the King Air would continue holding on Runway 4.

The NTSB's published report of the accident did not describe the Cherokee pilot's actions after he witnessed the accident. However, as reported in the NTSB Operations/Human Performance Group Chairman's report, after seeing the collision and the ensuing flames, the Cherokee pilot turned to passenger who was also a newly certified pilot and asked, "Did you see that? I don't want to get involved in this. I'm going to take off." The passenger reportedly replied, "You can't take off on that runway." To that, the Cherokee pilot stated, "I'll use this other runway," and they departed on Runway 36.

Survivability issues

Autopsy results indicated that occupants of both aircraft died from inhalation of smoke and toxic fumes, and not as a result of blunt trauma forces. The impact forces were at survivable levels, and should not have impeded the occupants' ability to evacuate. However, the Safety Board concluded that the rapid propagation and severity of the fire after impact prevented survivability for the King Air occupants.

The Safety Board concluded that the occupants of Beech 1900C were unable to evacuate because the cabin door could not be opened and the left overwing exit hatch was not opened. Although two pilots made multiple attempts to open the air stair from the outside by rotating the handle, the door could not be opened. One later described the door handle as being in the 6 o'clock position when he arrived at the burning aircraft. This position corresponds with the open position, suggesting that occupants inside the airplane also attempted to open it. The United Express first officer's body was found between the cabin door and the overwing exit, which led the NTSB to speculate that he likely attempted to open the door. Upon determining that the door could not be opened, in accordance with company emergency evacuation procedures, the first officer was likely proceeding to the left overwing exit when he was overcome by smoke.

The NTSB's materials lab examined the remains of the door and found all of the door locking cams in the locked position. The aircraft manufacturer indicated that when as little as one-quarter inch of slack is introduced into the door operating cable, the locking cams can be prevented from operating. The Safety Board surmised that the impact forces caused deformation of the door/frame system, which introduced slack into the door's cable system, effectively disabling the door.

"The Safety Board is concerned that even though the impact forces from the accident were so mild that both airplanes came to rest on their landing gear and the occupants of the Beech 1900C sustained little or no injuries as a result, those same forces were apparently sufficient [enough] to cause the Beech 1900C's air stair door to jam, preventing the occupants from using it to escape," stated the NTSB.

They concluded there were inadequacies in the FAA's certification requirements to show that external doors are reasonably free from jamming as a result of fuselage deformation.

Emergency response

The Safety Board remarked that the Quincy Fire Department that responded to the accident was ten miles away and did not arrive on scene until fourteen minutes after the accident. They contrasted that observation with FAR Part 139 requirements that certificated airports be capable of immediate response times of three minutes by on-site ARFF trucks equipped with extraction tools and staffed with properly trained firefighters. The Board noted that this FAA requirement applies only when the airport is being actively served by air carrier aircraft with seating capacity of more than thirty passengers.

The Quincy Airport ARFF truck was parked about 1,800 feet from the accident site, and if properly staffed, it could have reached the burning wreckage in less than one minute. "Fire fighters might have then have been able to extinguish or control the fire, thereby extending the survival time for at least some of the occupants of the Beech 1900C," stated the Safety Board. They further speculated that this extended survivability time could have allowed them more time to find, open, and escape through the overwing exit. "Accordingly, the Safety Board concludes that if on-airport ARFF protections had been required for this operation at Quincy Airport, lives may have been saved."

"The Safety Board concludes that although some communities may lack adequate funds to provide ARFF protection for small airports served by commuter airlines, commuter airline passengers deserve the same degree of protection from postcrash fires as air carrier passengers on aircraft with more than thirty passenger seats," concluded the NTSB. "Accordingly, the Safety Board believes that the FAA should develop ways to fund airports that are served by scheduled passenger operations on aircraft having ten or more passenger seats, and require these airports to ensure that ARFF units with trained personnel are available during commuter flight operations and are capable of timely response."

The probable cause

The NTSB determined that the probable cause of this accident was "the failure of the pilots in the King Air A90 to effectively monitor the common traffic advisory frequency or to properly scan for traffic, resulting in their commencing a takeoff roll when the Beech 1900C (United Express flight 5925) was landing on an intersecting runway."

The Safety Board cited a contributing cause as "the Cherokee pilot's interrupted radio transmission, which led to the Beech 1900C pilots' misunderstanding of the transmission as an indication from the King Air that it would not take off until after flight 5925 had cleared the runway. Contributing to the severity of the accident and loss of life were the lack of adequate aircraft rescue and firefighting services, and the failure of the air stair door on the Beech 1900C to open."

Recommendations

Because of this investigation, the NTSB made the following recommendations to the FAA:

- Reiterate to flight instructors the importance of emphasizing careful scanning techniques during pilot training and biennial flight reviews. (A-97-102)
- Issues surrounding the inoperable door: Evaluate the propensity of the Beech 1900C door/frame system to jam when it sustains minimal permanent door deformation and, based on the results of that evaluation, require appropriate design changes. Additionally, establish clear and specific methods for showing compliance with the freedom from jamming certification requirements. Finally, consider the circumstances of this accident when developing methods of showing compliance with freedom from jamming requirements, and determine whether it is feasible to require that doors be shown to be free from jamming after an impact of similar severity. (A-97-103, 104, and 105)
- Develop ways to fund airports that are served by scheduled passenger operations on aircraft having ten or more passenger seats, and require these airports to ensure that aircraft rescue and fire fighting units and trained personnel are available during commuter flight operations and are capable of timely response. (A-97-107)
- Add to the Safety Information Section of the FAA's Internet Home Page a list of airports that have scheduled air service but do not have aircraft rescue and fire fighting capabilities. (A-97-108)

- In addition, the NTSB reiterated to the FAA Safety Recommendation A-94-204, which stated that scheduled air carriers should only be permitted at airports certificated under the standards contained in FAR part 139, "Certification and operations: Land airports serving certain air carriers."

Industry action

The FAA continues to emphasize the importance of proper scanning techniques during pilot training and biennial flight reviews, and has recently published three journal articles and a video dealing with runway incursions.

An Airworthiness Directive (AD) was issued requiring highly visible markings on the outside of the Beechcraft door indicating the proper opening procedure. Upon review, the FAA found that there is no propensity for this door to jam, and that no design changes are necessary. Additionally, the FAA believes that "it is not feasible to show freedom from jamming for the door of the accident airplane because of the severity of the impact forces that were present in [the] accident." No changes are planned to Part 23 design standards, but wording changes may be implemented in guidance material used by aircraft certification offices in future projects.

Raytheon, the parent company of Beechcraft, submitted a petition for reconsideration to the NTSB regarding their conclusions in this accident, and the FAA is awaiting the Board's final resolution of the petition before finalizing their own response.

The FAA has initiated rulemaking to consider extending 14CFR 139 (airport standards) certification requirements to airports served by aircraft having ten to thirty-nine seats. They would then "consider requests on a case-by-case basis for Airport Improvement Program funds...for firefighting capital projects and equipment purchases at these airports."

Epilogue

This accident should send a powerful message to all aviators: something that you do as routinely as scanning for traffic or the way you communicate on the radio can have an critical impact on safety of flight. Radio communications must be conducted without ambiguity.

If an instruction, clearance, or other radio transmission does not sound quite right, it must be clarified.

Effective scanning for traffic means more than just looking out the window. The head and entire upper body must be moved in order to maximize the field of vision. The FAA's Advisory Circular (AC) 90-48C, "Pilots' role in collision avoidance," reminds pilots of the requirement to "move one's head in order to search around the physical obstructions, such as door and window posts. The doorpost can cover a considerable amount of sky, but a small head movement may uncover an area which might be concealing a threat." The AC also states, "Prior to taxiing onto a runway or landing area for takeoff, scan the approach areas for possible landing traffic by maneuvering the aircraft to provide a clear view of such areas. It is important that this be accomplished even though a taxi or takeoff clearance has been received."

Like so many accidents, this accident could have been avoided. Effective communication or proper scanning could have prevented this quandary at Quincy.

References and additional reading

Aarons, Richard N. 1997. Cause and Circumstance; Sad day at Quincy. *Business & Commercial Aviation*. February, 82.

Federal Aviation Administration (FAA). 1983. Advisory Circular 90-48C, *Pilots' role in collision avoidance*. Washington, D.C.: FAA.

Federal Aviation Administration (FAA). 1998. *Aeronautical Information Manual*. Change 2, January 28, 1999. Washington, D.C.: FAA.

Federal Aviation Administration (FAA). 1999. *NTSB Recommendations to FAA and FAA Responses Report*. <http://nasdac.faa.gov/>

Flight Safety Foundation. 1998. Landing aircraft collides during rollout with aircraft taking off on intersecting runway. *Accident Prevention*. January, 1-16.

National Transportation Safety Board (NTSB). 1997. *Aircraft accident report: United Express flight 5925 and Beechcraft King Air A90. Quincy Municipal Airport, Quincy, Illinois. November 19, 1996*. NTSB/AAR-97/04. Washington, D.C.: NTSB.

National Transportation Safety Board (NTSB). 1997. Public Docket. *United Express flight 5925 and Beechcraft King Air A90, Quincy, Illinois, November 19, 1996. DCA97MA009AB*. Washington, D.C.: NTSB.

Phillips, Edward H. 1997. Radio procedures key factor in NTSB Quincy investigation. *Aviation Week & Space Technology*. 7 April, 48.

12

Dangerous misconceptions

The legacy of Comair 3272

Operator: Comair Airlines

Aircraft Type: Embraer EMB-120RT

Location: Monroe, Michigan

Date: January 9, 1997

Safety of flight is dependent on the proper functioning of every component of the aviation system. Superb aircraft design, manufacture, operation, and maintenance typically couple with solid regulatory oversight to produce a very safe aviation network. Accidents rarely occur. Because of these successes, many of us tend to take this well-functioning system for granted. As passengers, we assume that the aircraft in which we fly have demonstrated compliance with all required design standards and adhere to appropriate regulations that are based on knowledge gained from decades of "hands-on" experience. As pilots, we rely on the fact that the performance parameters required of our aircraft have been demonstrated, and that the operational guidance and specifications issued by the manufacturer are accurate.

The tragic loss of Comair 3272 one wintry day early in 1997 demonstrates, however, that these really can be dangerous misconceptions: this story reminds us that in aviation there will always be new lessons to learn and erroneous assumptions to overcome.

Flight history and background

The day had started out pleasantly enough. A low-pressure area centered over Indiana brought typical winter weather to most of the northern midwest states. Surface temperatures hovered near freezing and light snow occasionally fell from the gray, overcast sky.

The Comair flight crew reported to work at the Cincinnati/Northern Kentucky International Airport (CVG) early on that Thursday morning to prepare for their flight. The first two legs, a quick round-trip to Dayton, Ohio (DAY), proceeded uneventfully, and by 12:40 in the afternoon they were back at CVG. The next leg of their trip was scheduled to the Detroit Metropolitan/Wayne County Airport (DTW), but would be delayed because the airplane scheduled for that flight segment, N265CA, was late arriving from Asheville, North Carolina.

Their aircraft, an Embraer EMB-120RT (see Fig. 10-1), was delivered to the airline in 1991 and had accumulated 12,751 flight hours. Pilots appreciated the sleek looks and favorable handling qualities of the twin-engine "Brasilia," named for the capital of the South American country in which it was designed and manufactured.

The captain of flight 3272 was forty-two years old and had more than 5,300 flight hours, with more that 1,000 as pilot-in-command of the EMB-120. He was extremely knowledgeable in Comair operational procedures, having been a CL-65 (a fifty-passenger twin engine jet aircraft built by Canadair and incorporated into the Comair fleet two years earlier) flight instructor and check airman. He had the reputation of being "very detail-oriented, professional and serious about his job." In addition to his normal flying duties, he assisted the Comair flight operations department in automating aircraft performance calculations and hoped to eventually direct the airline's performance engineering department when one was formed.

The other member of the cockpit team, the first officer, was twenty-nine. With 2,500 total flight hours, he had flown the Brasilia since being hired by the airline in October of 1995, and was also well respected within the Comair pilot group. That day, both pilots were eager to get the flight back on schedule, so they quickly serviced the airplane and reviewed the flight dispatch documents. With the preflight complete, flight 3272 departed the gate for DTW at 2:51 P.M., twenty-one minutes late. Because light snow was falling, they taxied to a designated deicing area, where an ethylene glycol/heated water solution was sprayed over the entire airplane. When that procedure was complete, they were cleared to the runway and were airborne at 3:09 P.M.

With the first officer at the controls, the takeoff and climbout were normal. Flying through turbulence, the crew requested and received clearance to cruise at flight level 210 (FL210, or 21,000 feet). The ride was smooth at that altitude, and the cruise portion of the flight was routine as they proceeded towards Detroit. The computerized flight plan indicated a brief forty-minute flying time, and at 3:39 P.M. a descent to 11,000 feet mean sea level (msl) was initiated. The DTW Automatic Terminal Information Service (ATIS) radio broadcast was received by the crew, with a recorded observation of one mile visibility, light snow, and low clouds. Additionally, several runways and taxiways were closed, and braking advisories (because of the slippery pavement conditions) and ground deice procedures (due to falling snow) were in effect.

After contacting DTW approach control, the crew was issued instructions to depart an arrival fix, MIZAR intersection, on a 050 degree (magnetic) heading and to expect vectors to the ILS (instrument landing system) approach to Runway 3 Right. They were also advised that the runway was slick; a DC-9 landing previously had classified the braking action as "poor."

A minute later, the air traffic controller radioed again. "Comair thirty-two seventy-two, maintain one niner zero knots...if unable advise." Another aircraft, an America West Airbus A-320, was also inbound to the same runway. Because that airplane was faster and had a more direct route to the final radar fix, the controller decided to sequence Cactus 50, the call sign for the A-320, ahead of the Comair flight. In order to maintain the required three-mile radar separation and to provide adequate wake vortex avoidance minima, Comair 3272 was slowed and vectored slightly away from the final approach course.

Fourteen seconds later, the Brasilia checked in with the next approach controller, advising level at 12,000 feet. Cleared to 7,000 feet, Comair 3272 passed over MIZAR intersection and turned to the northeast. Again, ATC issued a revised heading to the flight, 030 degrees, "for spacing." The captain acknowledged the clearance. A minute later, a right turn back to 055 degrees was issued, and the pilots complied with the instruction.

Approaching 8,000 feet in the descent, the first officer called for the descent checklist.

"Ice protection?" challenged the captain, reading the first item on the list.

"Windshield, props, standard seven," replied the first officer. His answer verified that the windshield and propeller anti-icing systems were activated, and that the "standard seven" anti-ice switches for the two angle-of-attack sensors, sideslip indicator, total air temperature sensor, and three pitot/static systems were on.

"Ignition?" continued the captain.

"Auto," was the response. The remaining six items were quickly completed.

The first officer then began his approach briefing, the first requirement of the approach checklist. "Okay...a thousand to go [the flight was just then descending through 8,000 feet for their assigned altitude of 7,000]...we're going to do an ILS to runway three right...it'll be a coupled approach...flaps twenty five...frequency is one one one point five...that's set...inbound course is zero three five." He continued with, "We're gonna intercept the top somewhere...ah...whatever altitude he gives us...twenty-seven hundred's the intercept to the glide slope." ATC then interrupted with instructions to turn further towards the right to a new heading of 070 degrees.

As the airplane turned, the first officer added power as the assigned altitude was reached. Finishing his briefing, he reminded the captain of the proper decision altitude for the approach, the missed approach procedures, and frequency selections. "Questions, comments?" he concluded.

"No questions...twenty-one, fourteen, and forty-three are your bugs," responded the captain. Small movable indices, or "bugs," on the airspeed indicator were then set to the approach reference speed, the takeoff safety speed, and the final segment airspeed, respectively. Just as the crew completed Autofeather and navigation radio checklist items, the controller radioed with additional instructions. "Comair thirty-two seventy-two, turn right to a heading of one four zero...reduce speed one seven zero..." The crew complied. Two items on the checklist remained to be completed, typically done when the aircraft was closer to landing: flight attendant alerting and selection of flaps to fifteen degrees.

"Comair thirty-two seventy-two Detroit approach, reduce speed to one seven zero and maintain six thousand...fly heading one four zero." The crew acknowledged, but since they had been steady on the assigned heading for over a minute, the first officer jokingly asked, "Wonder what plane he's looking at?"

"The one that's not going one four zero!" chortled the captain.

For the next thirty seconds the captain coordinated with the Comair station personnel via an alternate radio to complete passenger and fuel loading arrangements for the outbound flight, and received arrival gate information. "No changes while you were away," commented the first officer when informed that the exchange with Comair operations was complete.

A few seconds later, the final approach controller cleared the flight to 4,000 feet and the pilots started the descent. A discussion ensued between Cactus 50 and the controller regarding weather conditions at DTW.

"No [wind shear reports]...just slick runways and low visibilities," remarked the controller. "You'll pick up a headwind once you get down...oh...two thousand feet or so." Seven seconds later, he contacted flight 3272 again. "...Turn right heading one eight zero...reduce speed to one five zero." The crew acknowledged, but they were immediately assigned a new heading.

"...Now turn left heading zero nine zero, plan a vector across the localizer." The controller still needed to increase the spacing between Comair and the America West A320. His transmission alerted the crew that they could expect to pass through the localizer, the electronic runway centerline guidance, before being turned directly toward the airport.

"Heading zero niner zero, Comair thirty-two seventy-two," answered the captain. It would be the last radio transmission from the flight.

The airplane began a slow left turn to the newly assigned heading just as the autopilot began to level off at 4,000 feet. At 03:50:10 P.M. (EST), five seconds after the start of the turn, the plane had slowed to 156 knots and the left bank steepened to twenty-three degrees. The control wheel started to move back to the right to level the wings as commanded by the autopilot, but the left roll continued unabated.

A few seconds later, engine power started to increase, but the right engine torque value increased more quickly than did the left. At 03:54.17 the cockpit voice recorder (CVR) recorded a "significant decrease in background ambient noise," the control wheel moved even further to the right and the left bank continued to increase. The captain noticed an abnormal indication on his fast/slow indicator, an electronic presentation of aircraft airspeed relative to stall speed and commented, "looks like your low speed indicator." As the engines continued to increase in torque, the captain called "power!" Only a few seconds later, however, at 03:54.24, their speed had decreased to 146 knots and the airplane was rolling through forty-five degrees left bank. The stall warning "stickshaker" activated, followed immediately by autopilot disconnect and its distinctive "autopilot" aural warning.

Within two seconds, Comair 3272 was completely out of control. The airplane continued to roll until nearly inverted and the nose pitched down fifty degrees. The ground proximity warning system (GPWS) "bank angle" warning added to the noise of the stickshaker and other alerts. Exclamations of surprise and alarm from the pilots were among the last recorded sounds on the CVR. In a very steep nose down attitude, the Brasilia slammed into an open field next to a church campground in Monroe, Michigan, nineteen miles from its destination. All twenty-six passengers and three crewmembers were killed instantly.

The investigation and findings

The National Transportation Safety Board (NTSB) responded immediately to the accident site, with investigators from Washington, D.C., and from the regional offices in Chicago and Anchorage. Parties to the investigation included the Federal Aviation Administration (FAA), Comair, Inc., Empresa Brasileira de Aeronautica, Pratt and Whitney, the National Weather Service, the National Air Traffic Controllers Association (NATCA), and the Air Line Pilots Association (ALPA). In addition, an investigator from the Center for the Investigation and Prevention of Accidents (Brazil) provided assistance.

There was not much left of the airplane. Hitting at a high rate of speed, most of the wreckage was confined to an area around three impact crafters. The largest, about fifteen feet wide by twenty five feet long by four feet deep, contained what was left of the fuselage, while a smaller crater on each side held the embedded remains of the engines and propellers.

A post-impact fire burned most of the debris, but the Safety Board was able to determine that there was no evidence of an in-flight fire or preimpact damage or distress to either engine. Neither propeller, though extensively damaged, showed any sign of preexisting anomalies, nor was there any evidence that any aircraft system may have failed prior to impact. The flight data recorder (FDR) and CVR were located and sent to the NTSB labs in Washington, D.C., for readout. That data revealed that an in-flight upset was responsible for the accident, and after discounting propeller overspeed problems and other scenarios, attention soon focused on icing, wake turbulence, pilot actions, and aircraft flight controls.

Wake turbulence

In reviewing the air traffic control radar tapes, the Safety Board determined that the approach controllers followed all air traffic and wake turbulence separation rules. The ground tracks of Comair 3272 and Cactus 50 converged near the accident site, but analysis of the paths of the expected wake vortices and winds aloft indicted both horizontal and vertical separation existed, thus ruling out wake turbulence as a factor.

Icing

Engineers have been aware of the adverse aerodynamic effects of icing for many years—at least since the National Advisory Committee on Aeronautics (NACA) began studying the problem in the 1930s. That group determined that incredibly small surface roughness (.011-inch grains) covering only five to ten percent of the surface of the leading edge of a wing reduced the stall angle of attack (AOA) significantly, in some cases up to six degrees. In 1979 a Douglas Aircraft engineer wrote, "The effects of small amounts of surface roughness may not be particularly noticeable to a flight crew...[but] a 1.3Vs [1.3 times the stall speed of the aircraft] approach speed may have had the margin reduced [by the surface roughness] to 1.1Vs, leaving little actual stall margin for maneuvering or gust tolerance." He concluded by noting that accumulations equivalent to medium or coarse grade sandpaper can cause "a significant increase in stall speeds, leading to the possibility of a stall prior to the activation of stall warning." To provide an additional

margin of safety in these situations, some aircraft have a stall warning system that automatically reduces the angle of attack required
for stall warning activation when anti-icing systems are turned on.
However, the EMB-120 does not have this system.

There have been a significant number of icing-related aviation accidents in the last twenty years, resulting in special studies, Safety
Board reports and recommendations, Advisory Circulars (ACs), and
industry conferences and seminars. As pointed out by the NTSB, the
Safety Board has investigated about forty turboprop icing-related accidents since 1980. The most recent of these have highlighted the
hazards of operating in severe icing conditions and the aerodynamic
problems associated with ice accumulating on the tail of the aircraft,
or thin, rough ice accretions on airplane wings. As recently as 1996,
an FAA engineer suggested that the increasing number of roll excur-
sions due to in-flight icing was due to not only the increased utilization of turbo-propeller aircraft, but also the fact that these aircraft
"typically remain at lower altitudes for longer periods of time...and
are thus exposed to icing conditions for a greater percentage of their
flight time." Somewhat disturbingly, these remarks were made in the
wake of yet another turboprop icing accident.

EMB-120 development and certification

Manufactured by Embraer S/A in Brazil in December of 1991,
N265CA was purchased by Comair and flown to the United States
with a Brazilian Export Certificate of Airworthiness. All foreign manufacturers desiring U.S. airworthiness certification must undergo a
process similar to that to which U.S. builders must comply. The major difference is that the airworthiness authority from the country of
origin must provide "constant oversight of the...certification program
and participate in all certification flight tests," instead of the FAA providing that direction. Under the terms of the Bilateral Airworthiness
Agreement (BAA) between Brazil and the United States, the FAA recognized the Centro Tecnico Aerospacial of Brazil (CTA) as a competent airworthiness authority.

Embraer applied for a U.S. Type Certificate (TC) very shortly after fil-
ing a similar request for a Brazilian TC. As a result, the FAA was involved throughout the certification program, including "...a
preliminary board meeting...two interim board meetings and valida-
tion flight testing...[and a] final certificate board meeting." Also included were twelve "specialist" meetings between the two groups at

various times during the three-year development program. Since the airplane had been "examined, tested and found to conform to the type design approved under Type Certificate No. A31SO," a U.S. Standard Airworthiness Certificate was issued by the FAA for N265CA on February 20, 1992.

Icing certification

The certification requirements for flight into known icing conditions are contained in Part 25 of the Federal Aviation Regulations (FAR). To operate in continuous or intermittent maximum icing, "ice protection for various components of the airplane [must be] adequate, taking into account the various airplane configurations...[and] the airplane and its components must be flight tested in the various operational configurations, in measured natural atmospheric icing conditions."

Certification requirements in effect at the time stipulated a maximum intermittent water droplet size of fifty microns, and a maximum continuous water droplet size of forty microns. A micron is one thousandth of a millimeter, and for comparative purposes, the lead in a 0.5-mm mechanical pencil is ten times larger than the largest droplet defined in Part 25. Supercooled large droplets (SLD) have been shown to exist under some atmospheric conditions, however, and when present as freezing rain or drizzle can range in size from forty to four hundred microns in diameter.

Both CTA and FAA personnel participated in EMB-120 icing certification flight tests, with both natural icing conditions and simulated ice buildup on the wing. They found that the airplane demonstrated "satisfactory handling characteristics and no tendency for loss of control...." Airspeeds during the tests varied from 130 to 200 knots, icing accumulations were from one-quarter to three-quarters of an inch, and all aircraft configurations (gear and flap positions) were tested. Acceptable results were also obtained with three-inch "rams horn" ice shapes on the leading edge of the wing, representative of the rime ice that could accumulate during a one-hour hold in icing conditions. Finally, in several natural icing encounters during testing, conditions exceeded the Part 25 boundaries for continuous maximum and intermittent maximum icing conditions with no detrimental effect on aircraft handling.

The Brasilia is protected from ice accumulation and the resultant flight performance degradation by the following systems:

- Electrical anti-icing of the windshields, pitot tubes, static ports, and angle of attack (AOA), total air temperature, and sideslip sensors.

- Electrical deicing of the propeller blades.

- Pneumatic deicing, using inflatable "boots" to separate ice from the leading edges of the wings, the horizontal stabilizers, vertical stabilizer, and engine inlet and bypass ducts (see Fig. 12-1).

Engine bleed air is used to rapidly expand the boots, thus cracking the ice and separating it from the wing surface. Natural airflow past the leading edge then removes the ice. When deflated, the boots conform smoothly to the leading edge, maintaining the wing's airfoil shape. There were no preaccident maintenance difficulties with the anti-icing or deicing systems on Comair 3272, nor was there any evidence in the wreckage that the systems malfunctioned in flight.

Weather

National Center for Atmospheric Research (NCAR) Mesoscale Meteorological Study scientists assisted the NTSB in analyzing the various weather reports and pilot observations in the Detroit area at the time of the accident. The Safety Board found that "weather conditions were highly variable, but conducive to the formation of rime or mixed ice at various altitudes...rates...and amounts..." Furthermore, the research

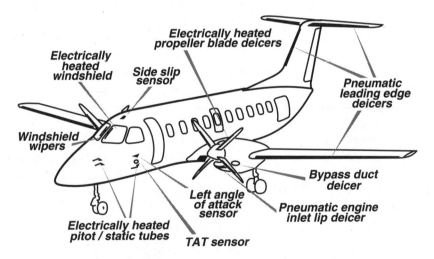

12-1 *EMB-120 ice protection systems.* . . . Source: NTSB

meteorologists concluded that "the evidence for the existence of icing conditions along the Comair 3272 flight track was strong and...that freezing drizzle-size droplets (forty to four hundred microns) might also have existed, especially along or near the cloud base...."

NTSB performance studies

Analysis of the FDR indicated to the Safety Board that a significant aerodynamic performance degradation was evident in the several minutes prior to Comair 3272's upset. To better understand that performance loss, an Embraer EMB-120 engineering simulator at Sao Jose de Campo, Brazil, was used to compare the data of the FDR with expected headings, altitudes, rates of climb and descent, bank angles, and airspeeds as programmed into the simulator's aerodynamic data bank.

To match the FDR data, Safety Board engineers determined that the accident aircraft was subjected to substantial lift degradation, drag increase, and changes in roll, pitch, and yaw moments. Analysis also indicated that something (presumably ice accumulation over time) caused a gradual elevator deflection, due to a reduction in the elevator and elevator tab aerodynamic efficiency. The gradual increase in drag was analytically applied at several intervals during flight 3272's descent profile, from 8,000 feet to 4,500 feet msl, by the use of "drag counts" (see Fig. 12-2). These units quantify the effect of drag on aircraft performance—the higher the count, the greater the aerodynamic effect, or performance loss. Using this technique, investigators were able to very closely match the engineering simulator flight profile with that of the accident flight. The Board was beginning to realize just how critical a role icing may have played in this tragedy.

Wind tunnel testing

The NTSB and the FAA conducted independent wind tunnel tests during the first year of the investigation. The Safety Board, in conjunction with scientists at the NASA-Lewis facility in Cleveland, Ohio, attempted to determine how much ice Comair 3272 may have actually accumulated and to identify the various effects different types of ice buildup would have on aircraft performance. The FAA contracted with the University of Illinois Urbana/Champaign (UIUC) to determine the implications of delaying activation of deicing boots, and what effect the residual ice or roughness remaining after completion of the boot's inflation cycle would have on aircraft performance.

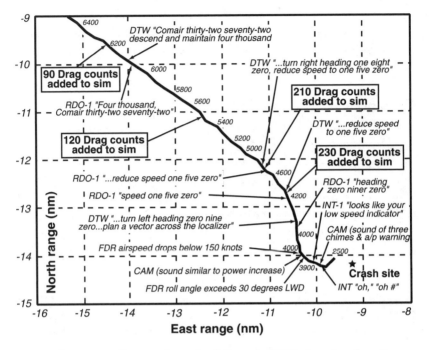

12-2 *Plan view of ground track of Comair 3272 showing locations of drag counts added in flight path simulation.* . . . Source: NTSB

The NASA study confirmed that in weather conditions similar to those encountered by Comair 3272, ice began to form quickly and initial accretions were rough, thin, and difficult to see. The ice was "extremely rough, like sandpaper," and was generally clearer than rime ice. During the testing, the Safety Board noted that the ice "would be difficult for pilots to perceive visually during flight, particularly in low light conditions (i.e., in clouds and precipitation, at dusk)."

Tests in the NASA Icing Research Tunnel (IRT) indicated that total ice accumulation varied, depending on droplet size, airfoil AOA, and total water content of the cloud. But total accumulation was determined not to be the most critical factor in aircraft performance. As the Board discovered, "a thin layer of rough ice with a small ice ridge...had a more adverse effect on lift, drag, and stall AOA than the three-inch ram's horn shapes...commonly used as a ' critical case' ice accretion scenario for FAA icing certification."

Results obtained from the FAA/UIUC testing confirmed that "potentially very hazardous aerodynamic degradation could occur before a pilot perceived ice accumulation on the airplane." Paradoxically, however, recommended procedures in place at the time of the accident suggested that pilots observe a measurable thickness of ice on the leading edge of a wing (one-quarter to one and three-quarters inches, depending on the aircraft manufacturer) before activating the boots. These practices were considered necessary because of concerns with "ice bridging," or the buildup of ice in the ridges of the deicing boot during periods of inflation, thus limiting the ability of the system remove the ice. However, according to evidence presented at the FAA/NASA Airplane Deicing Boot Ice Bridging Workshop in November of 1997, ice-bridging concerns in turbopropeller driven airplanes appeared to be the result of "myth or anecdotal incidents." The FAA's Environment Icing National Resource Specialist (NRS) stated that the safest practice would be for pilots to activate the leading edge deicing boots immediately upon entering icing conditions—"the earlier the better, cycle them and keep the wing contamination less than five percent, if possible."

History of related icing incidents

The NTSB had investigated an icing incident involving an EMB-120 near Pine Bluff, Arkansas, early in 1993. An upset and propeller blade failure had occurred after stickshaker activation and loss of roll control, due to undetected ice accumulation on the airplane. A Board search of NASA's Aviation Safety Reporting System (ASRS) found five additional EMB-120 icing events, with four of those documenting loss of control scenarios very similar to the Pine Bluff incident.

On October 31, 1994, an Avion de Transport Regional ATR-72 operated by Simmons Airlines crashed near Roselawn, Indiana, killing all sixty-eight people aboard. The aircraft, operating as American Eagle flight 4184, had been flying in a holding pattern while awaiting clearance to land at Chicago's O'Hare Airport. After a thorough investigation, the NTSB determined that the airplane had encountered severe icing, with SLD sizes between one hundred microns and four thousand microns in diameter, many times the size (forty to fifty microns in diameter) specified under FAR Part 25. The Board concluded that this accident "indicated that it was possible for pilots to unknowingly operate an airplane in icing conditions for which the airplane had not demonstrated satisfactory handling characteristics." As a result, the FAA requested that manufacturers of "regional transport-category airplanes with unpowered controls and pneumatic deicing boots" evaluate the handling qualities of those airplanes in SLD icing.

Embraer SLD tests

To comply with the FAA SLD request, Embraer utilized wind tunnel tests, flight simulator analysis, and flight testing in late 1995. The first series of wind tunnel tests were conducted to evaluate the effect of SLD on aileron hinge moment. It was found that a one-inch quarter-round strip of molding placed spanwise along the aft edge of the de-icing boot (to simulate an extreme ice buildup ridge on the wing) produced "a severe aerodynamic effect on the aileron (and general flight handling characteristics)." Further flight testing indicted that the airplane was fully controllable, but that "pilot-input" force required to maintain wings-level exceeded the sixty-pound maximum control wheel force stipulated for certification.

Embraer argued that the quarter round shape was "much more severe" than natural ice accumulation would be. In-flight icing tests were then performed to determine what the real shape of ice accretion on the leading edge of the wing would be, and what visual cues might allow pilots to better recognize freezing rain and freezing drizzle conditions.

During the tests, an airborne tanker aircraft generated droplets ranging from forty to seventy microns in diameter to simulate the SLD icing environment, while an EMB-120 flew behind the tanker and through the moisture cloud. The resultant wing ice shapes were less severe than the quarter-round molding, and control wheel forces in all cases remained less than the FAA maximum. Additionally, visual cues, including formation of ice behind the last inflatable tube on the boot and on the aft end of the propeller spinner were found to be accurate indications of flight in SLD conditions. In April of 1996, Embraer published an Operational Bulletin (OB No. 120-002/96) that provided all of the test results to EMB-120 operators and anyone else that was interested.

Pilot actions

The Safety Board identified three specific actions on the part of the Comair 3272 flight crew that warranted additional scrutiny: activation of aircraft anti-icing without activating pneumatic wing deicing, airspeed and flap configuration decisions, and use of the autopilot.

Use of deice/anti-ice equipment

As recorded on the CVR, the "standard seven" sensor heat, windshield heat, and propeller deicing were all activated as the airplane descended for the approach into Detroit. Guidance in the Comair Flight Standards Manual (FSM) for the EMB-120, however, cautioned against premature activation of the wing leading edge boots, which might result in "the ice forming the shape of an inflated deice boot, making further attempts to deice in flight impossible" (ice bridging). The Board noted that the pilots were aware they were in icing conditions, but even if they had observed any of the "thin, rough ice accretions that likely existed before the loss of control," they probably would have followed company guidance and not turned on the wing deicing system.

Airspeed/flap configuration

Comair pilots received operational guidance regarding EMB-120 normal and minimum airspeeds from several sources. Among those were the Embraer-published Aircraft Flight Manual (AFM), the Comair EMB-120 FSM, company bulletins, and cockpit "speed cards," small charts that provide normal approach and landing speeds for various flap configurations and landing weights. The NTSB noted that the company FSM did not contain specific minimum maneuvering airspeeds for flight in icing conditions. The speed cards indicated a normal minimum maneuvering airspeed of 147 knots (at their landing weight of 24,000 lbs., gear and flaps up) and a published stall speed of 114 knots in the same configuration.

A Comair interoffice memo circulated in December of 1995 advised pilots to maintain 160 knots in icing (gear and flaps retracted) and 170 knots if holding in icing. Portions of that memo were incorporated into Comair's EMB-120 FSM in October 1996. However, only the 170-knot holding speed was discussed, with no reference to descent or cruise speeds to be used in icing conditions. The NTSB found that "the language used, the different airspeeds and criteria contained in the guidance, Comair's methods of distribution and the company's failure to incorporate the guidance as a formal, permanent revision to the FSM might have caused pilots to be uncertain of the appropriate airspeeds for their circumstances."

Use of the autopilot

The NTSB determined that the automatic actions of the autopilot in the fifteen seconds prior to disengagement could have warned the pilots of the significant airplane performance degradation. The clues, though "subtle and short-lived," included movement of the control wheel to the right while the flight instruments indicated a continuing left bank. The sudden and unexpected disengagement of the autopilot allowed the ailerons to move rapidly in the left wing down direction, causing the airplane to roll to a nearly inverted attitude. The Board believed that had either pilot had his hand on the control wheel, recognized the airplane's loss of performance, and disengaged the autopilot, the event might have been more controllable.

The Safety Board concluded that "had the pilots been flying the airplane manually (without the autopilot engaged), they likely would have noted the increased RWD [right wing down] control wheel force needed to maintain the desired left bank, become aware of the airplane's altered performance characteristics, and increased their airspeed or otherwise altered their flight situation to avoid the loss of control." Additionally, this aircraft had a bank angle autopilot disconnect setting of forty-five degrees. If that limit had been set at a point corresponding to the initial autopilot bank angle command limit (about twenty-five degrees), investigators believe the crew would have had sufficient time to react to the increasing roll and recover the aircraft.

FAA oversight

In the analysis section of the final report, the NTSB was sharply critical of FAA actions in several important areas of regulatory oversight.

Icing certification

Reviewing all the test data, investigators were satisfied that the EMB-120 had met or exceeded all required icing certification requirements under Part 25, including the buildup of a three-inch "ram's horn" of ice on the leading edge of the wing. But the Board was distressed that there was no requirement for Embraer to demonstrate the airplane's performance after accumulations of other types of ice that could be encountered within the appendix C weather envelope. This would include thin, sandpaper-like ice with a small ice ridge just aft of the leading edge. The Board was also concerned that other potentially hazardous ice shapes, as yet unidentified, may exist, partly as a result of long-term FAA inaction in revising the 1950s-era Part 25 appendix C icing certification requirements.

Draft FAA report on EMB-120 icing concerns

The FAA was aware of the history of EMB-120 icing problems and had even presented a summary of the other six known events at an FAA/industry meeting in November of 1995. An FAA engineer reviewed these incidents in a draft report that detailed their concerns with the airplane's roll behavior after ice buildup, the high drag induced by ice accretions so small as to be unnoticeable by the flight crew, inadequate stall warning and stall speed margins in icing conditions, and problems associated with the use of the autopilot in icing. As Board Member George Black stated at the hearing, "Someone at the FAA knew. Looking back at that document, written almost a year before to the day, it certainly looks prophetic."

Revision 43 to the EMB-120 AFM

The FAA took no direct action in relation to the above-mentioned concerns, except to approve an Embraer-proposed, CTA-approved revision to the Airplane Flight Manual (revision number 43) that stipulated that the deice boots on the Brasilia should be activated as soon as icing conditions were encountered. According to the Safety Board, however, Aircraft Certification Office (ACO) personnel of the FAA apparently "did not accept the draft report's [other] conclusions, which recognized that pilots would not activate the boots if they did not recognize ice accumulation, that an engaged autopilot masked the tactile clues of icing, and that under these conditions, the flight-crew could also be deprived of adequate stall warning."

Furthermore, the FAA Principal Operations Inspector (POI) assigned to Comair and responsible for all aspects of the airline's operations and manuals was not aware of the Embraer Operations Bulletin OB No. 120-002/96 detailing the results of the SLD tanker testing. As such, he was unaware of the extensive rationale behind revision 43, which would have incorporated significant in-flight icing procedural changes to the EMB-120 AFM. Therefore, he did not require that revision 43 be incorporated into the Comair FSM or company procedures.

History of limited FAA action

The NTSB pointed out that many of the issues raised during this investigation were identified by the Safety Board as early as 1981. At that time, a study on icing avoidance and protection called for better FAA evaluation of individual aircraft performance in icing conditions, review and revision of Part 25 appendix C requirements, and establishment of standardized icing certification procedures.

The NTSB believed that FAA action in all of these areas was sporadic and ineffective. "Had the FAA adequately responded to the Safety Board's 1981 icing recommendation, the earlier accidents or the concerns expressed in its own staff's draft report on the EMB-120 [icing problems]...this accident would likely have been avoided." The Safety Board further noted that "The failure of the FAA to promptly and systematically address these certification and operational issues [autopilot use in icing, stall warning/protection system adjustment for icing, the effects of thin, rough ice and SLD, etc.] resulted in the pilots of Comair 3272 being in a situation in which they lacked sufficient tools...and information...to operate safely."

Conclusions and probable causes

The final report lists thirty-four findings in addition to the following probable cause. "The National Transportation Safety Board determines that the probable cause of this accident was the FAA's failure to establish adequate aircraft certification standards for flight in icing conditions, the FAA's failure to ensure that a Centro Technico Aeroespacial/FAA-approved procedure for the accident airplane's deice system operation was implemented by U.S.-based air carriers, and the FAA's failure to require the establishment of adequate minimum airspeeds for icing conditions, which led to the loss of control when the airplane accumulated a thin, rough accretion of ice on its lifting surfaces."

"Contributing to the accident were the flightcrew's decision to operate in icing conditions near the lower margin of the operating airspeed envelope (with flaps retracted), and Comair's failure to establish and adequately disseminate unambiguous minimum airspeed values for flap configurations and for flight in icing conditions."

Industry action prior to publication of the final report

One result of the 1994 Roselawn accident was the issuance of two recommendations to the FAA that called for revisions to Parts 23 and 25 icing criteria (A-96-54) and to icing certification testing regulations (A-96-56). The FAA responded by organizing an Aviation Rulemaking Advisory Committee (ARAC) tasked with "developing certification criteria for the safe operation of airplanes in icing conditions that are not covered by the current certification [appendix C] envelope." If funds could be found, additional research efforts into SLD icing could also be initiated. From the ARAC the Ice Protection Harmonization Working Group was established, tasked with generating research data to more closely define the SLD and mixed-icing weather environments.

The Safety Board acknowledged the efforts of the ARAC and various Working Groups to improve icing certification and aircraft system requirements, and urged the FAA to accelerate "research, development, and implementation of revisions to the icing certification testing regulations..." To that end, the Board reiterated Safety Recommendations A-96-54 and A-96-56 in the Comair 3272 accident report.

On March 13, 1997, the FAA held a meeting to review and discuss EMB-120 icing-related events, with most of the parties to the Comair investigation present. Three significant conclusions were reached:

1. Because visual cues alone are not sufficient, more precise guidance needs to be given to flightcrews as to when to activate deicing systems;

2. Activation of the boots at the first signs of ice accumulation should be mandatory;

3. Minimum speeds in various configurations and revised stall warning system parameters in icing should be considered.

These issues were addressed in Notice of Proposed Rulemaking (NPRM) 97-NM-46-AD and Airworthiness Directive (AD) 97-26-06, which required installation of ice detection systems on EMB-120 airplanes, adjustment of deice boot operational procedures, and identification of required minimum airspeeds. In May of 1997, to support the intent of the NPRM and to accelerate the regulatory process, the NTSB issued four Urgent Safety Recommendations. They recommended that the FAA:

- Require that all air carriers publish minimum airspeeds for all configurations in the EMB-120 operating manuals. (A-97-31)

- Ensure that all air carrier EMB-120 deicing procedures are consistent with the Embraer AFM. (A-97-32)

- Direct all POIs to ensure that all EMB-120 air carriers emphasize the importance of recognition of icing conditions and proper in-flight deicing procedures to flight crews. (A-97-33)

- Require all EMB-120 aircraft to be equipped with an automatic ice detection/crew alerting system. (A-97-34)

These recommendations were implemented through the FAA's issuance of the AD (effective date January 23, 1998) and a related

Flight Standards Information Bulletin (FSIB) on August 31, 1998. In part, the AD required EMB-120 flight crews to "activate the leading edge ice protection system at the first sign of ice formation...and maintain a minimum airspeed...of 160 knots." Additionally, operators were required to install an airframe ice detector on all EMB-120 aircraft by October of 1998.

Recommendations

Of the twenty-one new recommendations issued by the NTSB, two were directed to NASA and the remaining nineteen were to the Federal Aviation Administration.

The more noteworthy recommendations to the FAA included:

- Amend the definition of "trace ice" to preclude the impression that it is not hazardous. (A-98-88)

- Require POIs to discuss AFM revisions with their carriers and encourage the thorough dissemination of important operational information to the carrier's pilots. (A-98-89)

- In conjunction with other aviation organizations, organize and implement industrywide training on the hazards of icing and the importance of early boot activation, conduct research into the hazards of in-flight icing, develop and implement appropriate revisions to the icing certification regulations, and actively pursue the development of effective aircraft ice protection and detection systems. (A-98-90, 92, 93, and 99)

- Require manufacturers and operators of modern turbo-propeller aircraft to review and revise their manuals and training programs to emphasize the importance of early boot activation, and to provide stall warning systems that provide a cockpit warning before the onset of stall when operating in icing conditions. (A-98-91 and 96)

- Require manufacturers to provide minimum maneuvering airspeed information for all configurations and conditions, and to incorporate logic into all new and existing transport-category airplanes with autopilots to warn of bank or pitch angles that exceed the autopilot's command limits. (A-98-94 and 98)

- Require operators to incorporate the manufacturer's minimum maneuvering airspeeds into their manuals and training programs in a clear and concise manner, and to require turbopropeller pilots to fly the airplane manually when they activate the anti-ice systems. (A-98-95 and 97)

- Once the revised icing certification criteria are complete, review the certification of all turbopropeller aircraft currently operated and take action as required to ensure that these airplanes meet the revised standards. (A-98-100)
- Ensure that all FAA flight standards personnel are informed about manufacturer's operational bulletins and revisions, including all background material and justification. (A-98-103)

The two recommendations to NASA (A-98-107 and 108) concerned flightcrew training issues and were virtually identical to A-98-90 and A-98-92.

Industry action

The FAA agreed to remove the reference to "nonhazardous" in relation to trace ice as found in Order 7110.10, Flight Services, the Aeronautical Information Manual (AIM) and the Pilot/Controller glossary. This action was to be completed by July 15, 1999. Additionally, POIs were directed to encourage the establishment of a reliable "delivery" system for aircraft manufacturer's bulletins and AFM revisions within each airline. The action plan calls for a maximum of fifteen days from the date of bulletin issuance until receipt by the airline, with another fifteen-day limit for FAA approval. A flight standards handbook bulletin addressing this issue was to be published by the summer of 1999.

Another flight standards bulletin was issued that clarified the term "current" as it applied to company manuals, and emphasizes the importance of up-to-date information in the AFM, FSM and training programs. An electronic database listing all manufacturer's operations bulletins and AFM revisions will be established in mid 1999. FAA inspectors will use the database to evaluate and verify the operator's implementation of the material into their own manuals and training programs.

An FAA-sponsored In-Flight Icing Conference was held February 2-4, 1999. Attending were representatives from manufacturers, pilot and industry associations, the Safety Board and FAA personnel. Various working groups addressed ice protection, ice bridging, training, autopilot operation, weather, manuals and information distribution. According to the FAA, "products resulting from this conference will be used for possible changes to pilot information, training and icing regulations."

Late in 1998, letters were sent from the FAA to all manufacturers of turbopropeller transport-category aircraft requiring data that would show that their aircraft exhibit safe operating characteristics with ice accumulations on protected surfaces. The government has also published NPRMs for ADs requiring that "deicing systems be activated at the first indication of icing conditions and operated thereafter so as to minimize ice accretion" on fifteen different types of turboprop aircraft.

A working group formed as part of the FAA In-Flight Aircraft Icing Plan (April 1997) is reviewing guidance material related to the "identification and evaluation of critical ice accretions during the certification process." Cochaired by the FAA and NASA (Lewis Research Center), a report from the group is expected late in 1999. It is anticipated that the additional research necessary to accomplish the goal will take approximately five years.

One of the other working groups formed under the ARAC, the Flight Test Harmonization Working Group, has proposed a regulatory amendment to Appendix C to redefine aircraft 1-g stall speeds. Coordinated with an FAA sponsored change to the regulation, the goal would be published AFM operating speeds that "provide adequate maneuvering capability...with or without ice accretion, in all configurations and phases of flight." This project is scheduled for completion in December, 2000. This same group is also working on a better definition and understanding of wing ice shapes and the susceptibility of certain airplanes to tailplane stall, as well as developing criteria for more appropriate stall warning system performance requirements for Part 25 aircraft. Once those standards are finalized, the FAA will consider applying them to Part 23 commuter-category airplanes.

The Ice Protection Harmonization Working Group, yet another collection of government and industry representatives formed under the ARAC, has been tasked with developing SLD-type icing certification guidance. The group initially met in February of 1999 and their work is ongoing.

The FAA has determined that there are times when use of the autopilot during icing conditions is highly desirable or even necessary. They include flight through turbulence, while executing an instrument approach and during periods of high workload and pilot fatigue. After discussions with attendees at the February conference, it was decided that no restrictions on the use of the autopilot during flight in icing conditions were necessary. Revisions to the existing autopilot certification criteria for transport-category aircraft are currently under review, and changes, including autopilot function alerting features, may be included in policy to be issued as early as the fall of 1999.

Epilogue

This accident, like so many others, could have, and should have, been avoided. It was even prophesied—a year earlier the FAA had identified problems specific to the EMB-120 when operating in icing conditions. "Unofficial" recommendations stemming from that report, including several echoed by the NTSB as a result of other similar icing incidents, went unheeded. A false sense of security lulled the regulators, and to some extent industry, into inaction. Opportunities, and lives, were lost.

Fortunately, action is finally being taken. The ARACs, working groups, and the FAA will eventually find common ground and generate meaningful changes. But will it be enough? And what if the process takes too long? As Board Member John Hammerschmidt stated, "Just a little bit of ice can create a great deal of trouble for an airplane, and I hope we won't have to learn this lesson the hard way again."

References and additional reading

Federal Aviation Administration. 1999. *NTSB Recommendations to FAA and FAA Responses Report.* <http://nasdac.faa.gov/>

Evans, David, Managing Editor. 1998. Ice kills; pace of safety fixes glacial, Safety Board finds. *Air Safety Week.* 12h(35).

McKenna, James T. 1998. Comair probe points to certification flaws. *Aviation Week & Space Technology.* 24 August, 40.

National Transportation Safety Board. 1998. *Aircraft Accident Report: In-Flight Icing Encounter and Uncontrolled Collision with Terrain. Comair Flight 3272, Embraer EMB-120RT, N265CA, Monroe, Michigan, January 9, 1997. NTSB/AAR-98/04.* Washington, D.C.: NTSB.

National Transportation Safety Board. 1999. Public Docket. *Comair Flight 3272, Embraer EMB-120RT, N265CA, Monroe, Michigan, January 9, 1997. DCA97MA017.* Washington, D.C.: NTSB.

Part Three

Military Accidents

13

A missed approach

The fatal flight of Commerce Secretary Ron Brown

Operator: United States Air Force

Aircraft Type: CT-43 (Boeing 737-200)

Location: Dubrovnik, Croatia

Date: April 3, 1996

Most professional aviators would probably agree that the military pilot has one of the most exciting jobs in the sky. The tasks can be challenging, demanding total concentration and the utmost skill. In some squadrons, missions can vary from day to day; unfamiliar terrain and airports are the norm and operational flexibility is absolutely necessary to "get the job done." The military aviator usually has the opportunity to fly many different types of aircraft in his or her career, frequently in highly charged, complex, and interesting environments.

By contrast, the typical airline pilot's job could easily be seen by some as being a little boring! In late 1925, while surveying the new Chicago-St. Louis mail route for the Robertson Aircraft Company, even Charles Lindbergh complained of the "monotony" of scheduled line flying. In today's airline environment, pilots accumulate vast amounts of flight time but in only a few types of aircraft, often with minimal differences between them. The same routes are flown repeatedly, the same airports visited again and again, and the fundamental mission of safe, efficient, reliable, and comfortable transportation never changes. Standardization in every facet of flight operations is expected by management and instructors alike, and is usually attained.

That's one of the primary reasons why our commercial skies are so safe. Rapidly changing schedules and itineraries, unfamiliar airports, inappropriate supervisory pressures, poor flight planning and preparation, crew fatigue, and incomplete flightcrew training are not normally tolerated in civil aviation, nor should they be in the military sector. Any one of these conditions can, by itself, create an unsafe situation. When combined they virtually guarantee disaster, as evidenced one dreary day on a remote, fog-shrouded hillside in Croatia.

Flight history and background

The 76th Airlift Squadron (AS) of the 86th Airlift Wing (AW), US Air Force, had really been busy, flying its CT-43A, three C-20s (Gulfstream III), and nine C-21s (Learjet 35) almost nonstop. Based at Ramstein AFB in Frankfurt, Germany, the squadron's aircrews had flown eighty-four sorties in a sixty-day period in the spring of 1996, including trips for the Secretary of Defense and the First Lady, Ms. Hillary Clinton. So the operational tempo was already at a peak when another "distinguished visitor" (DV) transport request came in. Commerce Secretary Ron Brown and a delegation of U.S. industry executives would be visiting various locations in Bosnia-Herzegovina and Croatia on a three-day fact-finding mission in an attempt to develop closer economic ties with various Balkan countries recovering from the Bosnian civil war. The CT-43A aircraft of the 76th seemed the perfect plane for the demanding assignment. Basically just a military version of the tremendously popular Boeing 737-200 (see Fig. 13-1), the airplane was one of two in the Air Force that had been converted to passenger transport from its previous use as a navigational trainer. Outfitted with fifty-three seats and staffed with a flight crew of two pilots, a flight mechanic, and three in-flight passenger service specialists, aircraft #31149 was a "public use aircraft" of the U.S. government and was not required to be equipped with either a flight data recorder (FDR) or cockpit voice recorder (CVR).

Two of the squadron's best aviators were assigned the flight. The copilot was in the process of completing his upgrade to aircraft commander and had almost 3,000 hours of flight time, with 1,700 hours in the CT-43A. The pilot-in-command was an instructor and evaluator pilot in the aircraft, with 590 hours in the airplane and 3,000 hours total time. Although off to a rough start with the squadron in 1994 when his original commander refused to upgrade him, the pilot later received "squadron pilot of the quarter" honors and had re-

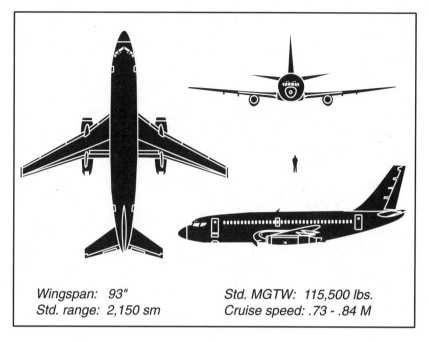

Wingspan: 93"
Std. range: 2,150 sm

Std. MGTW: 115,500 lbs.
Cruise speed: .73 - .84 M

13-1 *Boeing CT-43 (B737-200).*

cently found his career moving forward in high gear. He also enjoyed a solid reputation within the unit, being particularly respected for his thorough mission planning and crew briefings.

It was normal for all flight planning duties to be completed solely at the squadron level, so on April 1, 1996, two days prior to the flight leaving, the pilots prepared for the upcoming mission. Their call sign would be "IFO 21." But even as they were completing the task, their itinerary changed for the first time. At 7:45 P.M. the same day, change two was sent to the squadron, which incorporated a Dubrovnik stop in the schedule. There is no record that the pilots were made aware of the Dubrovnik stop until the following day, after all preliminary preparations for the flight were complete. The city of Dubrovnik and its airport lie at the extreme southern end of a narrow band of Croatia, wedged between the Adriatic Sea and the Dinaric Alps. Neither pilot had ever been there.

The co-pilot requested yet another change to the schedule the next morning, opting for a return to Zagreb while Secretary Brown and his party were meeting with officials in Sarajevo on Thursday, April 4th, the second day of the trip. The co-pilot also asked the squadron planner to work up flight plans for the Dubrovnik segments at the same time. That evening, the crew rode on another aircraft to Zagreb, the originating station for IFO 21, in order to be in position to start their flight the following day.

The pilots arrived at the hotel in Zagreb at 3:00 P.M. and agreed to meet in the lobby at 3:30 the next morning for their departure. But at 10:00 that night, the co-pilot was still concerned about the upcoming mission. He called Ramstein AFB and received a verbal summary of yet another change, number four. A hard copy of the new change was faxed to the hotel for his review, but only the cover sheet was received. The latest revision altered the flying schedule for the very next day, adding two additional flight segments. The Secretary's party had a full day of meetings planned in Tuzla; because there was insufficient ramp space to allow the aircraft to be parked at the Tuzla airport for the required seven hours, the plane would fly to Split, refuel, and return later in the day to pick up the delegates. They would then continue to Dubrovnik (see Fig. 13-2).

Two hours after receiving this information, the co-pilot was awakened by a phone call from a fellow squadron member. This pilot had

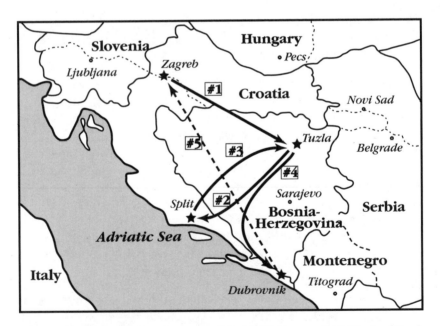

13-2 *Map of IFO 21 itinerary, April 3, 1996. Segment #5 (dashed line) not completed...* Source: U.S. Air Force

just flown the CT-43A aircraft in from Cairo, and wanted to deliver some personal clothing items to his friend. Additionally, the Cairo pilot was aware of the changes to IFO 21's itinerary, so when he stopped by the co-pilot's hotel room he also brought the latest mission change, planning data, and Dubrovnik instrument approach chart for review. By then it was 1:00 in the morning.

The latest weather reports for Tuzla, Dubrovnik, and Zagreb were faxed to the hotel from squadron headquarters at 4:03 A.M. It is unknown if the crew received them before departing for the airport, since their planned hotel departure was thirty minutes earlier. At 4:40 A.M., after arriving at the airport, the crew set up a phone patch through Thule Airways and a high-frequency radio link to contact U.S. Air Force Europe (USAFE) Meteorological Service (METRO) to obtain the weather for Tuzla and Zagreb, their first destination and alternate, respectively. The crew then filed five international flight plans, one for each flight segment that day. The flight took off from Zagreb at 6:24 A.M., twenty minutes later than anticipated due to the late arrival of the Commerce Department group.

Arriving in the Tuzla area fifty-one minutes later, IFO 21 was cleared for the Hi-Tacan approach to Runway 9. Shortly thereafter the approach controller notified the flight that it was "well left of the final approach course," to which the crew replied that they were correcting back. The crew also requested and received approval for a 360-degree turn in order to "lose a couple thousand feet."

After landing and disembarking all passengers, the flight took off for Split at 7:52 A.M., thirty-two minutes late. That segment and their return to Tuzla proceeded uneventfully, but because the flight crew had planned the use of a navigational corridor through Bosnia-Herzegovina that was not open, the rerouting caused an additional eleven minutes of unanticipated flight time. Once back at Tuzla, two extra passengers, a Croatian interpreter and Croatian photographer, were added to the official delegation, bringing the total number of passengers to twenty-nine. After everyone was aboard, IFO 21 finally departed Tuzla for Dubrovnik at 1:55 in the afternoon.

Deviating around thunderstorms after departure, they were instructed to contact "MAGIC," one of two NATO E-3 Airborne Early Warning (AEW) aircraft on station in the area to provide threat detection and high-altitude air traffic control (ATC) services to aircraft in Bosnia-Herzegovina airspace. A radio contact from IFO 21 to the operations center at Ramstein AFB confirmed that there were no new changes to the mission, but a caution was issued for possible fog at the Dubrovnik airport at the time of their arrival. An official weather report was then obtained from METRO, describing the current Dubrovnik conditions as "wind 110 degrees at 12 knots, ceiling 500 feet broken, 2,000 feet overcast, 8,000 feet overcast, altimeter 29.85, temperature 52 degrees [Fahrenheit]..." The visibility was reported as five miles, with rain in the area.

Soon, another navigational planning oversight became evident. Three corridors were approved for use through Bosnia-Herzegovinia airspace, each open for only a specific time "window." IFO 21 initially proceeded along the BEAR route, but again their planned route was not scheduled to be open at that hour. When MAGIC informed the flight that they were flying out of approved airspace, they asked for and received a vector to intercept an authorized airway. This change added another fifteen minutes to their flying time, delaying their arrival into Dubrovnik by over twenty minutes.

Upon reaching the Split VOR (Very high frequency Omnidirectional radio Range) and while communicating with Zagreb Center, IFO 21 was cleared to descend to FL 140 (14,000 feet). A normal descent was started and shortly thereafter a clearance was received to FL 100 (10,000 feet). Once level, air traffic control responsibility was transferred to the Dubrovnik Approach/Tower, a nonradar facility that relied on aircraft position reporting to maintain traffic separation.

The crew requested a lower altitude, but because of their proximity to another aircraft, descent clearance was delayed until they were about twenty-seven miles from the airport. Flying directly to the Kolocep (KLP) nondirectional beacon (NDB), later analysis indicated that IFO 21 flew at an airspeed of about 250 knots, most likely at idle thrust and with the landing gear up. Once level, the aircraft decelerated. The crew was preparing for the NDB Approach to Runway 12 at Cilipi Airport, and KLP was the final fix for the approach (see Fig. 13-3). The procedure required that the CV NDB, a different beacon, be used to identify the missed approach point of the approach.

Three minutes before crossing the KLP beacon inbound, the flight crew was contacted by a pilot on the ground in Dubrovnik and asked to communicate with him on another frequency. This pilot had landed his Challenger jet only an hour earlier, transporting the U.S. Ambassador to Croatia and the Croatian Prime Minister to

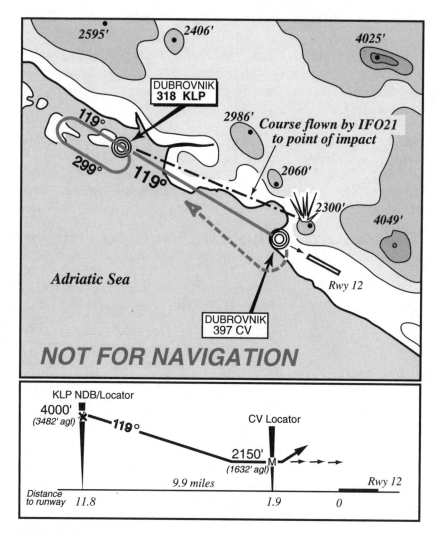

13-3 *NDB Runway 12 approach at Cilipi airport and flight path of IFO 21...*.Source: U.S. Air Force

Dubrovnik to meet with the Commerce Department delegation. An IFO 21 crewmember, probably the flight mechanic, established contact with the other aircraft using the second onboard VHF radio. Information was passed to the crew that the weather had deteriorated and was at minimums for the approach, and that an alternate circling procedure was available to Runway 30. It was also relayed that if a missed approach was necessary, IFO 21 should proceed back up the coast to Split, where the meeting would be held later in the day.

At 2:52 P.M., the crew radioed Dubrovnik tower that their position was sixteen miles from the airport, whereupon they were cleared to 4,000 feet and told to report crossing the KLP beacon. They began the descent and one minute later crossed KLP at 4,100 feet at an airspeed too high for final landing configuration (gear down and flaps thirty degrees) as required by the Air Force flight manual. Analysis of tracking data from the E-3 and Zagreb radars would later suggest that the actual indicated airspeed was between 209 and 228 knots. Weather at Dubrovnik at the time IFO 21 crossed KLP was reported as a broken cloud ceiling at 500 feet above ground level (agl) with an overcast at 2,000 feet mean sea level (msl), wind from 120 degrees at 12 knots, and five miles visibility.

No approach clearance had yet been issued to the flight, but instead of entering the holding pattern as depicted on the chart and required by international convention, IFO 21 initiated the Runway 12 NDB approach. "We're inside the locator [KLP], inbound," radioed the co-pilot. It would be the last transmission from the flight.

"Cleared for the approach..." responded Dubrovnik tower.

Within the next ten miles, IFO 21 would have to descend about 2,000 feet, lower the landing gear, slow the aircraft to allow positioning of the wing flaps to thirty degrees, complete all checklists, track the published outbound bearing from KLP, and determine the missed approach point. Things would be happening fast, but considering a normal approach speed of about two-and-a-half miles per minute, it could all be done. At four miles from the airport (two miles from the missed approach point, or MAP) the aircraft was level at the minimum descent altitude (MDA), gear and flaps were down, and the flight path and airspeed were stable.

But the airplane had not flown directly toward the runway, instead proceeding along a path several degrees to the east. At 2:57 in the afternoon, four minutes after beginning the approach, IFO 21 slammed into a rocky mountaintop known locally as "Strazisce" or St. John's Hill. Initial impact was only ninety feet below the summit. Located 1.7 nautical miles left of the extended runway centerline and 1.8 nautical miles north of the airport, the aircraft impacted in nearly level flight, at 2,172 feet msl altitude and at 140 knots. Most occupants were killed instantly, the force of the crash tearing the aircraft apart, scattering debris over an area 700 feet long by 460 feet wide. A fire broke out in the wreckage and burned for several hours.

Accident response

Realizing that IFO 21 had not landed at the anticipated time, the Dubrovnik tower made many unsuccessful attempts to contact the flight via the radio. Local police, military, and port authorities were quickly alerted. At 3:20 P.M., the NATO Combined Air Operations Center (CAOC) in Vincenza, Italy was notified. A French military unit at Ploce, Croatia, had the only locally available search-and-rescue helicopters, and their SA-330 "Pumas" were airborne by 4:55 P.M.

Initial search patterns were established over Kolocep Island and the surrounding water due to the anticipated approach path of the aircraft and a civilian report of a possible downed plane. Searching inland was impossible, because the low overcast clouds totally obscured the surrounding mountains. The CT-43A was equipped with a crash position indicator (CPI), an emergency radio beacon that would automatically transmit a signal in the case of an accident to aid in locating the aircraft. This device was similar to the civilian emergency locator transmitter (ELT) required on all U.S. general aviation aircraft. The range of the beacon was at least fifty miles, but it wasn't until 5:35 P.M. that the French helicopters received a very weak signal. Twenty minutes later a slightly stronger signal was detected, but no directional information could be ascertained.

Two U.S. Air Force MH-53J helicopters and an MC-130 tanker aircraft were dispatched from Brindisi, Italy, about an hour's flight time away. Arriving in the Dubrovnik area at 6:30 P.M., they had been given no information about the accident airplane's size, mission, or even the fact that it was conducting the published instrument approach. Both helicopters received strong enough signals to allow "direction finding" equipment to steer them toward the source of the signal, but they were unable to fly over the mountains due to the extremely poor weather conditions. Communications between the various rescue aircraft and Croatian air traffic controllers were difficult due to language limitations and ATC task saturation, at times hindering the passing of critical information.

At about the same time, local police received a call from a villager who had seen the accident site on top of a nearby mountain. This Croatian civilian had heard the sound of an airplane and an explosion, but because visibility was so poor, he did not see the wreckage for several hours. Having no phone, he then began the thirty-minute drive down the mountain to reach a phone and call for help.

By 7:40 P.M., almost five hours after the crash, five police officers finally arrived at the accident site. They initially found four bodies near the intact tail section and started looking for survivors. The rest of the airplane was so badly fragmented and burned, however, that they believed there could be no one left alive. Shortly thereafter, the U.S. helicopters landed at the airport amidst a torrential rainstorm, refueled, and prepared to depart for the crash site. The low cloud ceilings and restricted visibility prevented the crews from even seeing the bases of the mountains, let alone the upper slopes, so flying to the site was still impossible.

Within hours, ninety additional rescue personnel arrived at the site via ground transportation and a laborious trek across the mountain. At 9:30 P.M., one of the Air Force crewmembers on board the accident aircraft, an in-flight service specialist, made a sound. She had been found in the tail section under debris an hour earlier, but was believed deceased. A weak pulse was discovered, but the injuries to her spine, leg, and internal organs were severe. First aid was administered, and plans were made to move her off the mountain as quickly as possible. Airborne evacuation would have been best, but low clouds and restricted visibility still prevented the helicopters from flying to the site. Since immediate medical attention was necessary, the survivor was transported down the mountain by stretcher to a waiting ambulance. But the ordeal proved too much; at 11:55 P.M., a Croatian physician in the ambulance en route to the Dubrovnik Hospital pronounced her dead.

The first Americans on scene arrived at 1:45 in the morning, eleven hours after the accident. Two pararescuemen and a combat controller, all members of the Special Tactics Squad, were dropped in by rope from a U.S. helicopter, and by 4:00 A.M. many other U.S. military personnel had arrived via ground transportation. After a communications link and a base camp were established, a full-scale search for human remains began. Within twenty-four hours, all bodies had been removed to the temporary morgue at the airport. Autopsy results later confirmed that all twenty-nine passengers and five crewmembers died as the result of blunt force trauma, and the remaining crewmember succumbed to thermal inhalation injuries.

The investigation and findings

On April 5, 1996, the Commander U.S. Air Force Europe (USAFE) appointed Brigadier General Charles H. Coolidge, Jr., as President of the

Air Force Accident Investigation Board (AIB) tasked with determining the causes of the fatal crash of IFO 21. Other Air Force personnel and technical advisors from the U.S. National Transportation Safety Board, U.S. Federal Aviation Administration, Boeing, and Pratt & Whitney joined the investigative effort.

The Board was established under the provisions of AFI (Air Force Instruction) 51-503 for the purpose of finding and preserving evidence to use in claims, litigation, disciplinary actions, adverse administrative proceedings, and other purposes (see additional discussion, Chapter 14). The results of this investigation would be made public, whereas the parallel safety inquiry conducted under AFI 91-204 would remain confidential and for Air Force use only.

The aircraft

Earliest scrutiny focused on the demolished remains of aircraft #31149. All major structural components were found in the main debris field along a magnetic heading of 120 degrees. Investigators determined that there was "no preimpact failure or separation of aircraft structure," and because of the high decelerative forces, the accident was not survivable. The cockpit had been consumed by fire, but a complete teardown of the recoverable instrumentation provided investigators with the following information:

- One attitude direction indicator (ADI)—right wing five degrees low.
- Pilot's horizontal situation indicator (HSI)—course bearing between 115 and 119 degrees. The heading "bug" was set at 116 degrees, and the course select window, which is commonly used to display desired bearing, indicated 117 degrees.
- Pilot's bearing distance heading indicator (BDHI)—course bearing between 115 and 119 degrees.
- One radio magnetic indicator (RMI)—front glass, bezel, switch knobs, compass card and bearing pointers missing, needle position determined to be at either the 115 or 295 degree position.
- One inertial navigation system (INS)—operated for one hour, twelve minutes before impact; geographical coordinates were recovered and indicated that the equipment was operating within performance specifications. One control

display unit (CDU) display selector switch for one INS was set to the "cross-track/track angle error" position; the other CDU was set to the "heading/drift angle" position.

- One automatic direction finder (ADF)—the tuning synchro indicated a tuned frequency of 316 kHz (KLP frequency is 318 kHz), and the output synchro to the pilot's RMI indicated that the needle was pointing to 174 degrees. It was not determined if the difference in frequency selected was intentional or the result of the impact, but investigators believed it would not have significantly affected reception of the KLP NDB for the flight's navigational purposes.

- Attitude heading reference system (AHARS) and INS compass adapter, the unit that converts digital inputs to analog outputs for the pilot's HSI, RMI, and BDHI and the co-pilot's BDHI and autopilot—aircraft heading between 116.1 and 116.4 degrees.

Both engines were determined to be producing relatively high power at impact, with no indications of preimpact malfunction, bird ingestion, or thrust reverser deployment. The AIB determined that the high power setting could indicate that the crew was either initiating a missed approach, responding to a visual sighting of the rising terrain, or had been alerted by activation of the ground-proximity warning system (GPWS). Later analysis by the manufacturer, however, indicated that the GPWS would not have sounded because the flight path and configuration of the aircraft would not have triggered any of the preprogrammed alert modes of the system.

The crash position indicator (CPI) emergency beacon was found only partially deployed, remaining attached to the vertical stabilizer of the aircraft. Its signals, however, were blocked by the steep mountains to the south and southwest and the vertical section of stabilizer to the north and northeast. The battery pack was not damaged by the impact but was depleted, indicating normal activation.

Review of the airplane's maintenance records and work performed by Boeing contractors at Ramstein AFB found no discrepancies relevant to the accident. All maintenance personnel were properly certified, and interviews with managers at the base revealed "a high level of attention to detail and sound maintenance practices," according to the AIB.

The pilots

The Air Force AIB found all flight crewmembers fully qualified for the flying duties being performed at the time. Neither pilot had ever flown the Dubrovnik Runway 12 NDB approach before, but both had demonstrated proficiency performing other NDB approaches several times in the preceding two years. Human factors specialists noted no problems with flightcrew diet, sleep cycles, or personal stress, and postmortem toxicology analysis was "negative for medication, illicit drugs, or alcohol." Additionally, neither pilot had any abnormal preexisting medical condition that would have affected performance.

There was no indication that the aircraft commander had any interruptions during his preflight rest period in Zagreb. The co-pilot, however, had at least two interruptions, one for the phone call to Ramstein operations control and another to meet with the Cairo pilot. At the time of the accident, the crew was eleven and a half hours into a thirteen-and-a-half-hour duty day.

Weather

A widespread low-pressure system was responsible for low-level clouds, rain, and isolated thunderstorms throughout the Mediterranean region. Lightning activity to the north and south of the Dubrovnik airport confirmed that thunderstorms were in the area, and reports from eyewitnesses ranged from "light rain" to a storm some locals called "the worst in years." However, there were no reports or indications of any low-level wind shear activity.

Weather reports and forecasts issued to IFO 21 were timely and accurate, but in the hours after the accident, conditions at Dubrovnik deteriorated significantly. By 4:00 P.M., the wind had increased to sixteen knots, visibility was reduced to about one-half mile in rain, and a ceiling of only 300 feet was reported. These conditions remained for several hours.

Navigational aids and facilities

Various warring factions had frequently fought for control of the Cilipi Airport during the period 1992-1995, and during that time most navigational aids, including the Instrument Landing System (ILS), VOR, and a third NDB were stolen. The two remaining NDBs provided only one nonprecision instrument approach to the airport. Receiving both the KLP and CV beacons was necessary to complete the procedure, as KLP provided directional information to the runway, while CV was the only official way to determine the missed approach point (MAP). Since tuning one receiver to two different frequencies was impossible and changing frequencies in the middle of an approach was a violation of regulations, two separate airborne ADF receivers were required to fly this approach. IFO 21 had only one.

The Board considered the possibility that the crew may have attempted to identify the MAP by timing their arrival over the point based on anticipated ground speed between the final approach fix and the MAP. In that case, a reasonable time estimate would be about four minutes. The pilot's clock, as found in the wreckage, indicated an elapsed time of five minutes and fifty seconds. Assuming that the clock stopped upon impact, the AIB determined that the elapsed time feature must have been started prior to the final approach fix, thus not providing the information needed by the crew. Furthermore, the Board did not believe that the crew was attempting to identify the MAP by using an INS, because the position of the CDU display selector switches would not have allowed aircraft position information to be shown.

U.S. Federal Aviation Administration (FAA) flight-check inspectors verified the navigational accuracy of both beacons five days after the crash. Other electromagnetic effects, including high-intensity radiation fields (HIRF), coastal bending, electromagnetic interference (EMI), lightning, and signal reflection were investigated. None were found contributory to the accident. The AIB also looked into the possibility that the crew had tuned the wrong NDB, but determined that "The route of flight taken and the condition of instruments examined after the accident indicted KLP was successfully tuned on the ADF."

Design of the NDB Runway 12 approach

The International Civil Aviation Organization (ICAO) provides specific guidance to member nations for design and construction of instrument approach procedures. The Republic of Croatia's Air Traffic Service Authority claimed to have used these criteria when designing the NDB approach to Cilipi Airport, but the result was a nonstandard procedure. Investigators discovered problems with the length and definition of the final approach segment as well as irregularities in the identification of "controlling" obstacles. Those obstacles, mountains near the final approach path, should have been used in determining proper minimum altitudes and computation of the correct MDA.

The Croatian technician who developed the approach originally intended for the pilot to fly the inbound course to CV, not the outbound one from KLP. Secondly, a note on the chart should have been printed stating "timing not authorized for defining the missed approach point." Finally, when Air Force investigators computed an MDA based on appropriate ICAO standards, the correct minimum altitude was determined to be either 2,822 feet msl, using KLP as the primary navaid, or 2,592 feet msl using CV. Either minimum is significantly higher than the 2,150 feet msl that the crew of IFO 21 was using, and would have provided enough of a margin to clear the mountain (see Fig. 13-4).

Air Force supervisory responsibilities

During the investigation, the AIB uncovered several areas of regulatory noncompliance with the higher levels of command and control within the U.S. Air Force.

Terminal Instrument Procedures (TERPS) review

Air Force Instruction (AFI) 11-206 required that "any instrument approach procedure not published in a U.S. Department of Defense (DOD) or National Oceanic and Atmospheric Administration (NOAA) flight information publication be reviewed by the major command terminal instrument procedures (TERPs) specialist before it can be flown by Air Force crews." A supplement to this instruction allowed the use of non-DOD approach charts without review only if the weather was better than a 1,500-foot ceiling and 3.1 statute miles visibility. The intent of this instruction was to assure that appropriate margins of safety were built into the procedures in much the same way that they are in the United States. However, this review for the NDB Runway 12 approach at Dubrovnik was never completed, and therefore the approach should never have been flown.

When this guidance became official in November of 1995, it was clear that it would have a tremendous impact on the operational capability of the entire 86th Airlift Wing, as they were expected to maintain continuous mission readiness to many underdeveloped countries throughout the European and African region. The USAFE Director of Operations assumed that because Jeppesen-Sanderson printed and distributed the procedure, it must be "safe." But as the Board found, even Jeppesen is careful to point out that they "do not review or approve the adequacy, reliability, accuracy, or safety of the approach procedures they publish." The AIB discovered that the authors of AFI

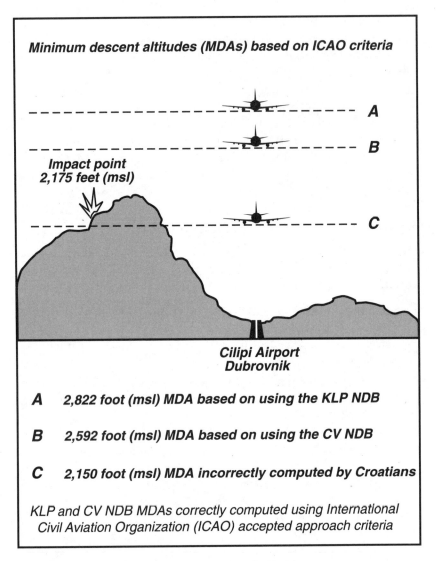

13-4 *Relative MDAs and impact point of IFO 21*...Source: U.S. Air Force

11-206 understood the implications of flying non-DOD reviewed approaches, but the staff at USAFE headquarters did not. Upon notification of the new requirement, the 86th Operational Group (OG) Commander immediately requested a blanket waiver from his superior, USAFE. The Air Force officer responsible for the request "believed he had obtained a verbal waiver...from AFFSA [Air Force Flight Standards Agency] during this conversation." The AFFSA officer testified later, however, that no verbal notification was given, nor would a verbal notification of a waiver ever be appropriate. But the misunderstanding allowed a "flightcrew information notice" to be distributed to all 86th OG pilots confirming that Jeppesen approaches could be flown as published without Air Force TERPS review.

On January 2, 1996, the official request for the waiver was processed and subsequently denied by Air Force headquarters. The e-mail message to USAFE read in part, "AFI 11-206 guidance on the use of Jeppesen approaches is sound...Proper approach development is a factor air crews 'take for granted' every time they fly an instrument approach." But the USAFE Director of Operations complained that "86th AW crews will have no authorized Jeppesen approaches to fly." So the 86th OG commander asked all squadron commanders to provide a list of approaches that would need the required TERPS review. But he concluded his request with the statement, "My view on this: Safety is not compromised if we continue flying ops normal until approaches have been reviewed—then we rescind [the flightcrew notice of waiver]."

Subsequent discussions between the OG commander, squadron commanders, and standardization / evaluation pilots produced agreement on this issue, and the notice was not rescinded. The AIB found that the 86th OG commander "understood at the time that the 86th AW was not following the letter of the [instruction]...[and he] believed that the entire chain of command knew of his action and was not directing him to act otherwise." The issue seemed closed until a representative of the 17th Air Force, the command to whom the 86th AW was responsible, learned that the 86th was still not complying with the AFI. He contacted his counterpart at the 86th and told him to "stop using Jeppesen approaches until reviewed." But the Board found that no action was taken; neither officer had followed up on the directive.

Crew Resource Management (CRM) training

The Accident Investigation Board recognized the importance of crew resource management, stating "Tenets taught in the resource management program are designed to help crews avoid mishaps like the one experienced by IFO 21 by improving skills for managing workload, air crew decision making and enhanced situational awareness." Two years prior to the accident, the Air Force issued AFI 36-2243, *Cockpit/Crew Resource Management*, requiring aircraft and mission-specific CRM training for all Air Force crewmembers. Although there was a direct responsibility to do so, USAFE had never developed a CRM program at all, leaving it up to the 76th Airlift Squadron to develop a program on its own. The pilots of IFO 21 had not attended this mandated training.

Operational pressure

In any DV mission, importance is placed on maintaining the expected schedule. As the AIB found, "external pressures to successfully fly the planned mission were present, but testimony revealed that [the crew of IFO 21] would have been resistant to this pressure and would not have allowed it to push them beyond what they believed to be safe limits." It was unknown exactly what pressures would have been generated by the Commerce Department group on the accident flight. However, in 1993 the Secretary of Commerce flew with the same squadron, and according to the Board, "scheduling difficulties were encountered, and a member of the Commerce party attempted to pressure the pilots."

Other issues

In the days immediately after the accident, a complete and accurate passenger manifest could not be located. According to the AIB, because specific directions in Air Force Instruction 11-206 had not been followed for IFO 21, a manifest had to be put together by the U.S. Embassy in Zagreb.

Another problem identified by the Board at the squadron level was the evolution of the CT-43A flight mechanic program. Originally staffed by "flying crew chiefs," it had turned into a "flight engineer" program with little or no formal training in flight duties. The sole remaining flight mechanic in the squadron "felt that he was not well trained and that he relied mostly on his on-the-job training."

Conclusions and probable causes

Upon concluding its investigation, the Accident Investigation Board published its 7,174-page report in twenty-two volumes. Areas that were investigated and found not be substantially contributory to the mishap included aircraft maintenance, structures and systems, crew qualifications, navigational facilities, and medical qualifications of the crew. The Board also stated that "although the weather at the time of the mishap required the aircrew to fly an instrument approach, the weather was not a substantially contributing factor to this mishap."

The AIB also found that:

- Air Force commanders at many levels "failed to comply with governing directives from higher authorities," particularly relating to approach review procedures contained in AFI 11-206.

- Air Force Command failed to provide theater-specific training to those crews routinely flying into underdeveloped airfields utilizing non-DOD instrument approach procedures. As the Board stated, "Proper training would have enabled the aircrew to recognize they could not fly the Dubrovnik approach with the navigational equipment on the aircraft. They should not have attempted to do so."

The Board also found that the flightcrew of IFO 21 made a number of critical errors that, when combined, were a cause of the mishap. These included:

- Inadequate mission planning, approach procedure review, and route selection. Four changes to the mission itinerary contributed to planning difficulties encountered by the crew.

- Initiating a rushed approach without proper clearance, leading to late configuration of the aircraft and higher than planned airspeeds.

- Distractions caused by the rushed approach, late configuration, and extraneous radio communications.

- Failure to properly identify the missed approach point. As the Board stated, "Had they executed a missed approach at the missed approach point, the aircraft would not have impacted the high terrain which was more than one nautical mile past the missed approach point."

Finally, investigators stated that the Dubrovnik NDB approach procedure was improperly designed. Applying correct obstacle clearance criteria would have resulted in a higher MDA, placing the aircraft "well above the point of impact, even though the aircraft flew 9 degrees off course."

As stated in the final "Statement of Opinion" written by Accident Investigation Board President Coolidge, "The CT-43A accident was caused by a failure of command, aircrew error and an improperly designed instrument approach procedure." He went on to say, "This mishap resulted from a combination of the three causes...any one of which, had it not existed, would have prevented the accident."

Epilogue

The highly contested waiver to AFI 11-206 was quietly and uncere-moniously rescinded by the 86th OG on April 4, 1996, one day after the accident, but one day too late for the crew and passengers aboard IFO 21. A special Air Force group, called a "tiger team," was established to inventory safety equipment onboard the service's transport aircraft. The team was tasked with performing a cost analy-sis for installation of CVRs, FDRs, Terminal Collision Avoidance Sys-tems (TCAS), and Global Positioning Systems (GPS) for those airplanes not so equipped.

Because of the investigation, two generals, the 86th AW commander and the 86th OG commander, were relieved of command. The US-AFE director of operations and the 86th AW vice-commander re-ceived letters of reprimand, and twelve other Air Force officers were disciplined.

Sure, after any major accident, there will always plenty of blame to go around. But it is cause, not blame, that investigators seek. In civil avia-tion, "tombstone" regulation often drives necessary safety improve-ments. But Air Force policy limits the flow of investigative information, and therefore it is difficult to know if the crash of IFO 21 and the sub-sequent investigation produced any real changes that might prevent similar tragedies in the future. One improvement that has been made— late in 1997 a Mobile Microwave Landing System (MMLS) was de-ployed on many C-130 and C-17 aircraft. This system can be set up by three people in just an hour, and would allow precision approaches and landings in adverse weather at "unprepared" airports.

Distinguished Visitor flights will always be a part of overseas diplomacy. But while a mission-oriented, "can-do" attitude is es-sential in armed conflict, it must yield to a philosophy of stan-dardization and safety—lessons so hard-learned in the commercial aviation environment.

References and additional reading

Aarons, Richard N. 1996. Cause and Circumstance: CFIT in Dubrovnik. *Business & Commercial Aviation*. February, 86.

Anonymous. 1996. U.S. Air Force deploys new microwave landing sys-tem. *Aerospace Daily*. Issue 36, 282.

Coolidge, Charles H., 1996. *AFI 51-503 Accident Investigation Report, CT-43A, 73-1149, April 3, 1996, Dubrovnik, Croatia.* United States Air Force.

Eurich, Heather J. 1996. USAF studies safety systems upgrades for its passenger airplanes. *Defense Daily.* Issue 13.

Flight Safety Foundation. 1996. Dubrovnik-bound flight crew's improperly flown nonprecision instrument approach results in controlled-flight-into-terrain accident. *Flight Safety Digest.* July, August, 1-25.

Hughes, D., Couvault, C., and Proctor, P. 1996. USAF, NTSB, Croatia probe 737 crash. *Aviation Week and Space Technology.* 8 April.

Kern, Tony. 1997. *Unforced Errors: A Case Study of Failed Discipline.* Colorado Springs: HQ USAFA/DFH.

National Transportation Safety Board. 1996. Public Docket. *United States Air Force, CT-43A, 73-1149, Dubrovnik, Croatia, April 3, 1996. DCA96RA46.* Washington, D.C.: NTSB.

Weiner, T., Bonner, B., Perlez, J. and Wald, M. 1996. A 737 below civilian safety standards. *New York Times Service, International Herald Tribune,* 29 April.

14

A lack of teamwork

HAVOC 58 impacts Sleeping Indian Mountain

Operator: United States Air Force

Aircraft Type: Lockheed C-130H "Hercules"

Location: Jackson Hole, Wyoming

Date: August 17, 1996

For over half a century, the U.S. Air Force has maintained a proud tradition of excellence. Whether defending the nation's interests or just projecting impressive military might to any point on the globe, success has been achieved through careful planning, precise operational execution, and teamwork.

The differences between a scheduled airline "flight" and a tactical military "mission" are usually significant and obvious. But when the only objective is the routine transportation of government personnel and equipment, disdainfully referred to as "trash hauling" by some military aviators, the distinctions between the two operating environments blur (see Chapter 13, "A missed approach"). Proper coordination, communication, and planning at all levels are essential. Effective mission support and appropriate prioritization of mission goals must be maintained. Without these prerequisites, the operating crew becomes the last link in the safety chain, with little margin for error. As the crew of HAVOC 58 demonstrated, any mistake at that point can be deadly.

Flight history and background

"Salt Lake Center, HAVOC five-eight has Jackson Hole in sight."

The C-130 descended quietly to 14,000 feet for the approach to the municipal airport serving the city of Jackson Hole, Wyoming, and the resort areas of the Teton National Forest. The night was impressively dark—no moon, few stars, and only occasionally did a lonely ground light twinkle through the scattered lower-level clouds. Virtually invisible to the busy flightcrew, the majestic peaks of the Rocky Mountains' Wind River Range loomed only a few thousand feet below the Hercules.

"HAVOC five-eight roger, cleared visual approach to the Jackson Airport, report your cancellation or arrival time this frequency, change to advisory frequency approved."

The tactical airlifter of choice for over forty years, the Lockheed C-130 is capable of hauling ninety-two combat troops or 30,000 pounds of support equipment into virtually any landing strip in the world. Its cavernous cargo hold is ten feet wide, nine feet high, and forty-one feet long. The twenty-three cockpit windows provide excellent visibility, and the expansive wing and tail surfaces allow unusually slow airspeeds for parachute drops and assault landings. In service with over sixty other air forces worldwide, the "Herk" has been pressed into service as a transport, aerial refueler, firefighter, gunship, hurricane hunter, and reconnaissance platform. It may well be the most versatile aircraft ever designed (see Fig. 14-1).

The pilot continued the descent and maneuvered the four-engine transport to align with Runway 18 at the Jackson Airport. Arriving from their home base of Dyess AFB, near Abilene, Texas, this was only the first leg of several scheduled for this Saturday night. The primary mission for this crew was a comprehensive JA/ATT (Joint Airborne/Air Transportability Training) exercise scheduled for the following Monday at Pope AFB, Fayetteville, North Carolina, and extensive planning had gone into preparing for the thirteen airdrops the crew would perform in just a six-day period. But a late-breaking order had arrived only three days earlier, advising the 39th Airlift Squadron of an additional mission: in support of the Presidential vacation in Wyoming, critical security equipment needed to be returned to the East Coast by the 18th of August. Since no other crews were available, HAVOC 58's flight schedule was altered. These Special Assignment Airlift Missions (SAAM) were routine for the squadron, and this one, code-named PHOENIX BANNER because of the presidential involvement, was typical. Abrupt changes to the itinerary and crews were not completely unexpected, but were sometimes difficult to accommodate.

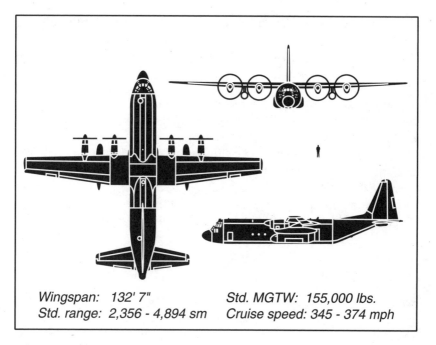

Wingspan: 132' 7"   Std. MGTW: 155,000 lbs.
Std. range: 2,356 - 4,894 sm   Cruise speed: 345 - 374 mph

14-1 *Lockheed C-130H.*

After a quick stop in Jackson Hole to load cargo, fuel, and a Secret Service agent, the flight would continue to John F. Kennedy Airport outside New York City. After safely depositing the Treasury Department equipment and passenger, HAVOC 58 was to reposition to Pope AFB, arriving Sunday morning. It was going to be a very long night.

Final checklists were completed and an uneventful landing was completed. "...HAVOC five-eight's on the ground at Jackson Hole, close out that leg of the flight plan, please sir," radioed the female Air Force co-pilot to Salt Lake Center.

"HAVOC five-eight roger, and I do have your flight plan to JFK."

"Outstanding," she said, "see you in a bit."

Taxiing to the south end of the ramp area, the C-130 was marshaled into a secure area reserved for the numerous military transports that had recently transited the field. By about 9:40 P.M., the engines had been shut down and most of the eight-member flight crew had migrated toward the pilot's lounge. The flight engineer, however, headed toward the airport manager's office in pursuit of fuel, and the navigator contacted the Casper Flight Service Station (FSS) for a thorough weather briefing.

At the aircraft, a bulletproof Suburban and six cargo pallets of Secret Service luggage and sensitive communication gear were loaded through the airplane's aft ramp/door. The exact composition of the shipment was classified, but its weight totaled 8,500 lbs. and was valued at $1.3 million (U.S.). The accompanying Secret Service agent also boarded for the flight to New York, and refueling was completed by 10:15 P.M. On the way back to the airplane, an airport employee asked the flight engineer if he was aware of the airport's noise abatement takeoff procedures. "Yes," was all he said. With servicing complete, final preparations were made for departure.

The airplane's cockpit voice recorder (CVR) captured the preflight checklist interphone conversation. "How do you want the navaids [onboard navigational radios] set up for departure, Pilot?" asked the navigator.

"...You tell me what makes sense, because I am not really sure of what the course is and all that stuff out here...I assume it will probably be set up Jackson Hole initially but I don't know what our course is defined off of," replied the aircraft commander.

The normal start sequence was interrupted by a manual shutdown of the number three engine, in response to a problem similar to that encountered earlier in the day at Dyess. The flight's crew chief, coordinating the start-up from outside the aircraft, commented to a nearby marshaler that it was not a mechanical problem, but a pneumatic one. With the number-three engine running, there was insufficient bleed air to start the other engines. After starting the left side engines, however, the other two started normally.

The engineer completed all takeoff computations and relayed the data to the pilot. The before taxi checklist was resumed. "Radios and navigation equipment?" challenged the copilot.

"What did you decide on, Nav? Just, uh...?" asked the aircraft commander.

"I've got Jackson in all around on the navaids," replied the navigator.

During the crew briefing two minutes later, the commander mentioned to the rest of the crew, "...once we get airborne and climbing away, we will probably end up setting ten-ten [power setting] for the cruise. But uh, just to get climbing away from all these mountains...and we'll pull that back after a bit." And a few seconds later, "...What do we have off the end of the runway there, Nav, do you know?"

"...It's pretty much a mountain and valley terrain, rising to seventy three hundred feet, about a thousand foot rise, ten miles south of the field."

"All right, now if I need to turn, I need to turn right, correct? Or excuse me, left?" asked the commander.

"Left turns," responded the navigator.

"Okay, because that mountain we came across is up that way. Okay, so it would be a left turn to climb, if we need it and we will try to do a left hand turn pretty quick after takeoff anyway."

"Yeah, we would like a pretty sharp left turn and a pretty rapid climb rate," answered the navigator.

"Okay," stated the aircraft commander. "Climb rate we'll have to work on, the left turn, I think we can do. Everything else remains as briefed. Anybody got any questions? The only thing is if we need to do an emergency return back here, Nav, I am going to need you pretty actively looking to see where terrain is."

"Okay."

"HAVOC five-eight," radioed Salt Lake Center, "you're cleared from the Jackson Airport to the JFK airport as filed, climb and maintain one six thousand, squawk [transponder code] six zero two seven...Hold for release and you can call me when you're at the end of the runway..."

"HAVOC five-eight copies that..."

The copilot then broadcast their intention to taxi to the end of the runway on the UNICOM (advisory) frequency used at the airport. "Roger," responded the radio operator in the airport operations office. "Preferred runway is one eight, no known traffic in the area. And are you aware of the noise abatement procedure?"

"Yes," answered the aircraft commander to the copilot. "HAVOC five-eight copies all and we are aware of the noise abatement procedure," radioed the copilot.

The ramp area to the north was heavily congested with aircraft; twenty minutes elapsed before the crew of HAVOC 58 had carefully picked their way out to the parallel taxiway. At one particularly tight spot, the right wingtip of the C-130 slowly passed over the tail of a small airplane parked on the ramp. Once clear, the crew chief quickly went aboard.

"So," commented the navigator, "we are looking at pretty much an immediate left turn after takeoff, Pilot, direct to the Boysen reservoir, uh, TACAN."

"Okay."

The taxi checklist continued. "Instruments?"

"Checked, Pilot." "Copilot."

"Nav select panels?"

"Uh, see here," queried the pilot. "You said it was going to be a left turn to what heading, Nav?"

"Uh, it'd be a left turn to zero eight zero."

"Left to zero eight zero," acknowledged the pilot. "That's going to be my terrain clearance heading too?"

"Yea, that's affirmative," the navigator replied. "The sooner we can turn, the better."

Before takeoff, checklists were completed. The latest weather and altimeter setting was received. But just as the crew was about to call for their Air Traffic Control (ATC) release, an inbound Skywest Airlines regional airliner radioed in, "...straight in runway three six, we're seventeen DME [distance] out."

"Oh (expletive deleted)! Three six, huh?" exclaimed the aircraft commander. Their planned takeoff on Runway 18 would put that traffic directly in front of them after departure and create a potentially hazardous situation. Just then another transmission was received, this one from a Delta Airlines jet also preparing for takeoff.

"Jackson traffic, Delta sixteen-ninety-one taxiing for departure, Jackson Hole, planning runway three six, copy inbound traffic."

Somewhat exasperated, HAVOC 58's pilot commented, "Ah, why is everybody going to [runway] three six? I thought they said preferred runway was one eight!"

"Yeah," answered his copilot, "I thought that is what he said too."

Realizing that the longer they waited the more difficult it would be to utilize Runway 18 for takeoff, the command pilot appropriated control of the radio from the copilot. Hoping to preempt any plan that the Delta airliner might have to take off ahead of them, he transmitted, "Jackson Hole traffic, HAVOC five eight, Charlie one-thirty is in the approach end of Runway one eight, holding for traffic." Frustrated, to his crew he commented, "Freakin' Delta airplane going up to three six...(expletive deleted)...let's give Center a call on that and...just let 'em know we are awaiting a release with an inbound airplane." The C-130 was equipped with only one VHF communications radio, necessitating rapid switching of frequencies and the inability to listen to both ATC and UNICOM at the same time.

"Salt Lake Center, HAVOC five-eight...at the approach end of one eight waiting for our release," radioed the copilot. ATC responded almost immediately.

"HAVOC five eight is released, be advised there is an aircraft inbound for landing at Jackson Hole...showing one one miles south of the field and showing out of one one thousand feet. You are released, climb and maintain flight level one niner zero [19,000 feet], you can report your departure on this frequency and the, uh, Skywest should be over there on UNICOM."

"Roger," answered the copilot, "thanks."

Planning to depart after Skywest landed, and anxious to go, the pilot commented, "This guy is slower than we are! As soon as he is on the ground and clear, we'll see if we can beat this Delta guy outta here."

Two minutes later, the commuter flight landed and rapidly cleared the runway. "All right," exclaimed the pilot, "let's go for it!"

With all the checklists complete, HAVOC 58 lifted off at 10:47:19 P.M. Twelve seconds later, a left turn to the east was started. "Zero eight zero, Pilot," reminded the navigator.

"Thanks, and Eng can you kill those flight deck domes, I don't think you got them all the way off." Told that they were already turned down, the pilot replied, "Are they? It's awful bright in here, maybe it's just me."

The after takeoff checks were done, and at 10:48:38 P.M. airborne contact was established with the Center. "Salt Lake Center, HAVOC five eight's passin' eight thousand for one nine thousand." Center acknowledged. "I'll tell you what," said the pilot, "let's keep one seventy in the climb for now."

"All right, Co, you ready to fly?" The aircraft commander had performed the takeoff, but it was his intention to let the less experienced copilot fly the remainder of the trip to New York. "...And copilot is ready to fly...I have the aircraft," she acknowledged.

Thirteen seconds later, the copilot was startled by a sudden indication on her radio altimeter, probably because of rapidly rising terrain below the aircraft. "My radar altimeter just died!" she exclaimed. Eerily prophetic, two seconds later all nine people aboard HAVOC 58 also died as the C-130 slammed into the side of Sleeping Indian Mountain, just a few hundred feet below the ridge line. A giant fireball erupted, visible for many miles on such a dark night, and the incredibly fragmented wreckage of the Lockheed Hercules continued to burn until daybreak.

Emergency response

By 2:00 A.M., Bridger-Teton National Forest (BTNF) rangers and the Jackson County Sheriff had organized an Incident Management team to find the accident site. Twenty-eight personnel from BTNF, Grand Teton National Park, the Teton County Sheriff's Department, and the U.S. Secret Service comprised the rescue team. Traveling on horseback and on foot, the group reached the grisly scene at about 4:30 A.M. After quickly determining that there were no survivors, the area was secured.

The site was located nine-and-a-half miles from the airport, in a desolate region of the Gros Ventre Wilderness Area of the National Forest (see Fig. 14-2). As evident in Figs. 14-3 and 14-4, the plane had burned a swath through the rugged underbrush for about a quarter of a mile, and the resultant fire had destroyed the wreckage.

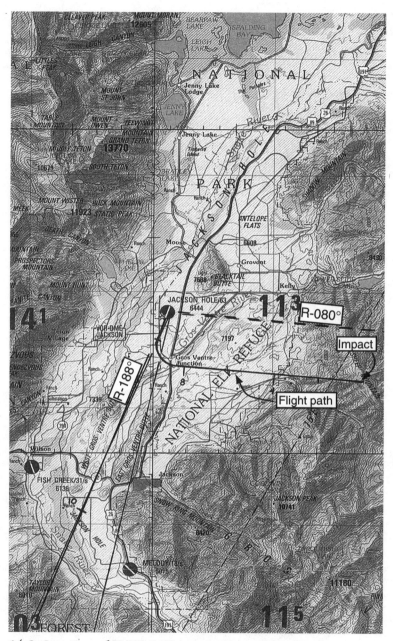

14-2 *Location of HAVOC 58 impact...*Source: U.S. Air Force

14-3 *View of HAVOC 58 wreckage, looking to west...*Courtesy U.S. Air Force

14-4 *View of HAVOC 58 wreckage, looking to south...*Courtesy U.S. Air Force

A photographer from the local newspaper and his friend trekked fifteen miles through desolate mountains to document the crash. Arriving at sunrise, they were met by a very surprised Teton County search and rescue team. For security purposes, they were kept several thousand yards from the debris, but no attempt was made to prevent their photographing the scene.

"Smoke was still rolling off [the wreckage]," the reporter said later. "It was pretty eerie." Eventually they were asked to leave, but were intercepted on their way back down the mountain by security officers. Their film was confiscated for "national security purposes," but was returned undamaged later that afternoon with no explanation.

The investigation and findings

Investigations of military aircraft accidents differ from their civilian equivalent in structure and function, but hopefully not in purpose. Upon notification of the Class A mishap (loss of aircraft or personnel), the Eighth Air Force Commander, Lieutenant General Philip Ford, appointed Colonel Robert Skolasky as the Accident Investigation Officer. Acting under the provisions of Air Force Instruction (AFI) 51-503, *Aircraft, Missile, Nuclear and Space Accident Investigations*, Colonel Skolasky then appointed technical and legal advisors to assist him in the inquiry. Additionally, pilot, navigator, flight surgeon, and flight engineer investigators participated as well (see additional discussion regarding the conduct of military accident investigations in Chapter 13, "A missed approach").

The aircraft

Careful documentation of the wreckage was begun. Investigators measured the characteristic ground scars caused by the rotating propeller blades. Knowing the climb power RPM setting and the fact that the C-130's Allison engines had four-bladed props, investigators computed the aircraft's ground speed to be 182 knots at impact. Nose landing gear structure and door parts were found furthest down the hill, with the heavier components, such as the engines, further along the debris path. High terrain on the left side of the airplane caused the left wingtip to strike the hillside first, leaving a noticeable ground scar and wing fragments early in the accident sequence. As the cargo floor disintegrated, the forward part of the fuselage tumbled, the wing center section separated, and all fuel tanks ruptured.

Detailed analysis of aircraft wreckage indicated that the engines were developing normal power and the flaps were retracted upon impact, with no evidence of any obvious flight control problems. Additional scrutiny of the flight data recorder (FDR) and CVR confirmed apparently normal aircraft operation prior to impact.

The Jackson Hole Airport

About twelve miles due north of Jackson, Wyoming, the Jackson Hole Airport is centered in the Snake River valley. The airport elevation is 6,444 feet mean sea level (msl), but immediately to the west, the Grand Teton mountain range towers to almost 14,000 feet. To the east lies Sleeping Indian Mountain, part of a north-south ridge rising over 10,000 feet.

Surrounded by the Bridger-Teton National Forest, the National Elk refuge, Grand Teton National Park, and the homesteads of local residents, aircraft noise had been an environmental issue of local concern. To alleviate the problem, noise abatement procedures were developed for the field that limited aircraft departures to specific ground tracks. A takeoff to the south was preferred to avoid the National Park. An immediate forty-five degree turn to the left after departure avoided houses off the end of the runway, but rising terrain prevented maintaining that heading for more than a few miles. Instrument Flight Rules (IFR) departure procedures dictated a left turn back over the airport while climbing to a minimum safe altitude before turning on course. These procedures were published in the Jeppesen approach charts for the field, were posted on the wall of the airport flight planning room, and were available as a handout to pilots at the airport operations desk.

According to investigators, military pilots did not usually have access to the Jeppesen charts, but relied on Flight Information Publication (FLIP) Terminal Procedures and Department of Defense (DOD) approach charts. Those documents, however, did not mention the approved noise abatement procedures.

Terrain avoidance procedures, including IFR departure routings, were described in detail in all the publications. DOD charts utilized a "T" within an inverted black triangle (see detail in each corner of Fig. 14-5) to call attention to a specific terminal departure procedure that exists at a particular airport and must be flown. That symbol is readily apparent on the chart and refers to Volume 3, page XX of the FLIP Low Altitude Publication. The procedure calls for a climb to 11,000 feet msl while tracking outbound on the JAC VOR 188-degree radial, then a climbing left turn direct to the VOR (see text, Fig. 14-5). Detailed in-

XX

GILLETTE, WY
GILLETTE-CAMPBELL CO   Rwy 21,34 600-1ᵃ
                                                        Rwy 16. 1100-2ᵃᵃ
ᵃOr standard with a minimum climb of 350ᶠ
NM to 5000ᶠ
ᵃᵃOr standard with a minimum climb of 300ᶠ
NM to 6000ᶠ
Rwy 3. 16 turn left. Rwy 21, 34 turn right. All
aircraft climb direct GCC VOR/DME. Aircraft
departing GCC VOR-DME R-283 CW through 230ᶜ
climb on course. Aircraft departing GCC VOR-
DME R-231 through 282 climb in GCC VOR/DME
holding pattern to 8000ᶠ before proceeding on
course. (HOLD-N RT, 164 inbound).

GRAND JUNCTION, CO
WALKER FIELD   Pwy 04 NA
                                    Rwy 11 2300-2ᵃ
                                    Rwy 22, 29 2600-2ᵃᵃ
ᵃ Or standard with a minimum climb gradient
of 290ᶠ/NM to 8000ᶠ
ᵃᵃ Or standard with a minimum climb gradient
of 300ᶠ/NM to 8000ᶠ.
Comply with SIDS or: Rwy 11, 22 climbing right
turn direct to JNC VORTAC. Rwy 29 climbing left
turn direct to JNC VORTAC. Aircraft departing
JNC R-221 CW R-060 depart on course. all other
aircraft continue climbing in the JNC holding
pattern(hold SW, LT, 052ᶠ inbound) to cross at or
above: R-061 CW R-130 9500ᶠ, R-131 CW R-220
10.500ᶠ

GREELEY, CO
GREELEY-WELD COUNTY, CO
Rwys 27. 35 turn right. Rwy 9. 17 turn left.
All aircraft climb direct GLL VORTAC. Aircraft
departing GLL R-301 CW R-212 climb on course.
All others continue climbing in GLL VORTAC
holding pattern (Hold NE, RT, 205ᵒ inbound)
to cross GIL VORTAC at or above: R-213 CW
R-257, 11,200; R-258 CW R-300, 5000ᶠ or comply
with SID or radar vectors

GREYBULL, WY
SOUTH BIG HORN COUNTY
All runways, for departures on GEY NDB
bearing 320ᵒ CW 150ᵒ climb in GEY NDB
holding pattern to 9000ᶠ before departing on
course. then continue climb to MEA or assigned
altitude. Departures GEY NDB bearings 150ᵒ
CW 320ᵒ climb on course.

JACKSON HOLE, WY   Rwy 18. 3700-3ᵃ
                                                Rwy 36. 3600-3ᵃ
ᵃOr standard with minimum climb of 270ᶠ/NM to
10,100ᶠ
ᵃᵃOr standard with minimum climb of 310ᶠ/NM to
8800ᶠ
Rwy 18. Climb to 11,000ᶠ via JAC R-188, then
climbing left turn direct JAC VOR/DME. Aircraft
departing JAC VOR-DME R-356 CW R-037ᶜ or
R-142 CW R-207 climb on course. All others
continue climb in JAC VOR/DME holding
pattern (hold S. RT. 006 inbound) to cross JAC
VOR-DME at or above R-038 CW R-141. 12,300ᶠ
R-208 CW R-279. 12,200; R-280 CW R-355. 15,000ᶠ
Rwy 36. Climb Rwy heading to 7300ᶠ then clim-
ing right turn to intercept JAC R-014 to NALSI
INT/JAC 17-3 DME. continue climb via DNW
R-267ᵒ to DNW VOR/DME. Aircraft departing
DNW VOR/DME R-281 CW R-016 and R-088
CW R-218 climb on course. All others continue
climb in DNW VOR/DME holding pattern (hold
W. RT, 087 inbound) to cross DNW VOR/DME
at or above: R-017 CW R-087. 12100; R-219
CW R-280 13700ᶠ

KINGMAN, AZ
                            Rwy 3, 17, 21, 35. 300-1
ᵃOr standard with minimum climb of 210ᶠ per
NM to 7000ᶠ.
Climb northbound on IGM R-010 until reaching
6000ᶠ. reverse course to the left and continue
climb to 9000ᶠ direct to IGM. If unable to cross
IGM at 9000ᶠ continue climb in holding pattern
SW of IGM LT, 027ᵒ inbound.

LAGUNA AAF, YUMA PROVING
GROUND, AZ
Rwy 6, 24, 36 turn left. Rwy 18 turn right.
climb to 3600ᶠ direct to 8ZA VORTAC.

LAMAR, CO
LAMAR MUNI
Rwys 8 and 36 turn left; Rwy 18 turn left/right.
Rwy 26 turn right; direct LAA VORTAC. Aircraft
departing LAA VORTAC R-048 CW R-118 climb
on course. All others continue climbing in LAA
VORTAC holding pattern (Hold N. RT. 169ᵒ
inbound) to 6000ᶠ before proceeding on course.

LARAMIE REGIONAL, WY
Rwys 12-21 turn right. Rwy 3, turn left. All
aircraft climb direct LAR VORTAC; continue climb
in holding pattern; (W. LT, 107ᵃ inbound) to cross
LAR VORTAC at or above: Wbnd V-4, 10,600ᶠ
All others cross LAR VORTAC at or above MEA
for direction of flight.

LAS CRUCES INTL, NM
Rwy 4-30, climb on rwy hdg to 5500ᶠ prior to
turning northbound.

14-5 *Jackson Hole FLIP departure procedures...*<small>Source: U.S. Air Force</small>

structions then follow for various airways and altitudes to be flown during the en-route climb phase of the flight.

One of the squadron's navigator instructors explained to investigators, however, that these procedures were rarely encountered. "I have never flown into any fields that had the 'T'," he said. "That's something you catch once a year in the navigator instrument re-fresher course. I don't know anybody that has ever used one."

The airport director later told investigators that prior to flying into his airport, "military folks call us [for departure procedure information] because of the en-route supplement instructions. It is not as clear or as amplified in the IFR en-route supplement as it is in Jeppesen. I would also say that seventy percent of military pilots call...during flight planning prior to arriving here." A squadron mission planner did call and speak to a Jackson Hole airport administrator, but only to verify that no special permissions were required for HAVOC 58. No operational issues were discussed. Investigators found no evidence that any crewmember had called the airport for information or had reviewed specific noise abatement or IFR departure procedures for the Jackson Hole Airport.

ASR/GDSS

To obtain operational information about any airport that might be included in a military itinerary, the mission planner or pilot would typically consult the Airport Suitability Report (ASR). This document provides basic airfield information including location, available facilities, runway weight-bearing capacity, and operational limitations for the 2,907 airports surveyed by the Air Mobility Command (AMC).

Another source of airport information was the Global Decision Support System (GDSS), a complex, unclassified computer system used by the Air Force for mission scheduling, flight following, and data retrieval. Initially designed in the 1980s, the network had been upgraded many times since. A revised "windows"-based system was distributed to all users in 1996, with the 39th squadron receiving theirs in mid-July. Administrative personnel that input and download the relevant data didn't "get the books" (operating instructions) until the end of July, however, and no formal training was offered until the end of August. One feature of the new software was the Airfield Database (AFD).

As designed, all of the information previously available to mission schedulers only through the ASRs could be instantly retrieved in the AFD. But the users told investigators that the system was more useful for "mission following" than for data retrieval. Operators continually had problems logging onto the network and the revised menu-driven interface tended to be confusing. These limitations prevented the system from being used to its full potential; the Deputy

Commander of the Operational Group stated that he was "unaware that we even use GDSS." None of the squadron planners interviewed knew that the AFD portion of the system was available for their use.

Investigators learned that at the time of the accident, the Airport Suitability Report for Jackson Hole indicated no special operational considerations for AMC aircraft. The computerized Airfield Database differed from the ASR, however, in reporting that the airport was suitable for "daylight operations only." Neither the flightcrew nor the planners were aware of the restriction recently placed on military operations at the airport.

The officer responsible for inserting the change explained that the airport had the appropriate lighting and approach facilities (a precision approach Instrument Landing System, or ILS) for mountainous terrain, so operational safety was not the overriding concern. But unknown obstacles existed in the proximity of the taxiways and runway, creating a potential wingtip clearance problem for another aircraft in the military transport fleet, the C-17. He acknowledged that if a waiver for nighttime operation had been requested for HAVOC 58, it would have been granted.

The squadron

During the spring and summer of 1996, all of the Air Force's Airlift Squadrons (AS) were heavily tasked. Ninety-six BANNER missions in August alone were scheduled, and routine training, normal assignments, and overseas deployments had to be completed as well. The 39th and 40th AS, both based at Dyess AFB and "across the street" from each other, were no exception. The 39th had been deployed to Southwest Asia and the 40th AS was due to relieve them, so the work schedule during the first few weeks of August was even more frenzied than usual.

Qualified crewmembers were in critically short supply. As one squadron operations officer told investigators, "We were severely limited [short of aircraft and crews] at that time, especially crew use, due to the deployment and some confusion with the Saudi swap out." Many C-130 crews were "backed up" and unavailable because of relocation requirements and crew rest conflicts. According to the Commanding Officer (CO) of the 39th AS, the overall experience level of the flightcrews was much less than in prior years, straining an already ambitious schedule.

"I feel like the tempo of operations is totally out of control," complained one instructor during the investigation. "I think it's really hard on everyone. I'm TDY [temporary duty, usually away from the home station] all the time. It is very difficult for us."

Administratively, the 39th AS was in some disarray. Investigators found that the squadron was significantly behind in processing OPRs and EPRs (required personnel performance reports) and the CO was "contemplating administrative reorganization" to address the paperwork and training problems. An internal squadron safety questionnaire distributed shortly before the accident (but not included as part of the official accident report) pointed out many of these issues. "People violating post-mission crew rest, the heavy ops tempo and the emphasis on nonflying duties" were cited by one pilot, creating a "generally callous attitude about safety."

Investigators uncovered other squadron-level problems. According to a flight engineer instructor, there was "not much emphasis here on the decreased capability of the aircraft operating out of high altitude airfields...because we don't deal with it that much...we really don't have a lot of mountains in the areas that we fly." Another operations officer referred to it as a "flat-area mentality," a safety issue that was known but not addressed.

In the midst of all this activity, the 39th AS was assigned the PHOENIX BANNER mission. No aircraft or flight engineers were available in the squadron, so both were "borrowed" from the 40th AS. Assembling the rest of the crew would prove difficult as well.

The flightcrew

The only navigator available to fly PHOENIX BANNER was brand new. Having just completed initial training, this assignment would be his first without an accompanying navigator instructor. Additionally, it was his first "off-station" (away from Dyess AFB) sortie. He was eager to do well, but confided in a former roommate that because of squadron staffing shortages he thought "he was moving a little bit too quickly on a few things." He had experienced some difficulty learning to interpret the terrain-mapping portion of the onboard radar, but otherwise was progressing normally. One of his instructors later told investigators that "for terrain avoidance, you have to play with it a year or more before you become proficient in identifying terrain...Reading the radar for weather is a little bit easier."

The navigator had been sent to Pope AFB earlier in the week for High Altitude, Low Opening (HALO) drop training, but that exercise had been canceled due to a mechanical problem with the aircraft. It was critical, therefore, that his HALO certification be completed as planned on Monday, August 19, for squadron participation in the subsequent JA/AAT exercise.

The aircraft commander was also the senior executive officer of the 39th squadron. His administrative duties required that he not be scheduled to fly at all unless absolutely necessary and only when specifically approved by the CO. In this case, though, no one else was available. He was also a C-130 instructor pilot, highly regarded by his peers. "One of the most conscientious, conservative, by the book kind of fliers I have ever known," according to another instructor.

The crew chief came with the airplane from the 40th AS, and the loadmasters were all assigned from the 39th AS. The schedulers were forced to go "purple," mixing the crew from different units, an undesirable, but not uncommon practice.

Mission planning

The addition of the BANNER assignment had stretched the mission to eight days, violating a squadron-imposed limit of seven-day trips. Originally, the mission had been scheduled to depart at 2:00 A.M. Saturday morning from Dyess AFB, allowing a morning takeoff from Jackson Hole. While the crew was in the operations office Friday morning, however, word came down from the White House Military Office that the time had slipped—the new plan was for a 4:00 P.M. Saturday afternoon takeoff and a late night departure out of Jackson Hole.

The team had prepared thoroughly for the JA/AAT exercise and had carefully reviewed the arrival and departure procedures for John F. Kennedy Airport in New York. The Jackson Hole portion of the trip, added just the day before, was reviewed for normal daytime operations. No pilot in the squadron had ever flown into the airport, nor had any nighttime military operations been conducted there in support of the presidential visit.

The navigator had been given all of the Pope AFB charts a few days earlier by an instructor and retrieved necessary ASRs and FLIPs from the duty desk. The crew then spent most of midday Friday planning the mission. The copilot noticed that a waiver for nighttime operations at Tri-State Ferguson Airport in Huntington, West Virginia (the last stop on the JA/AAT), was required, and one was obtained. Two navigator instructors passed through the planning room, but neither had time to assist the HAVOC 58's navigator—one had to give a check ride and the other needed to complete a postmission debrief. The crew then dispersed, not reuniting again until the next afternoon.

For some unexplained reason, the flight engineer reported to the aircraft early Saturday morning, mobility bag packed, ready for the trip. He left when told that the departure time had changed, but investigators were puzzled as to why he had not been notified earlier. Although his mandatory crew rest period had been interrupted, he reported as planned later in the day for HAVOC 58.

Conclusions

The AFI 51-503 Accident Investigation Report was published October 21, 1996, after a five-week investigation. The Investigating Officer found violations of regulations in five areas:

- Interruption of crew rest required under AFI 11-401, due to a lack of communication between the aircraft commander and the flight engineer.
- Failure of the crew to leave a copy of the DD Form 365-4, weight and balance, with personnel on the ground at Jackson Hole, Wyoming, as required by MCR 55-130.
- Assignment of the co-pilot to "nonflying or noncombat related activities" while in her first year of her initial operational assignment, in violation of MCI 11-C-130.
- Failure of the flight crew to "adequately prepare for and execute a departure in mountainous terrain." Specifically, the crew did not review published departure procedures, failed to cross-check the airborne radar presentations with flight instruments indications, and did not utilize all available navigation aids during the departure.
- Taxing over the tail of the smaller aircraft on the ramp violated MCR 55-130, which required a clearance of ten feet with wing walkers.

The probable cause

Colonel Skolasky's Statement of Opinion, as attached to the final report, stated, "The crew of HAVOC 58 failed to avoid the mountainous terrain ahead. They were complacent and not situationally aware of their proximity to that terrain. Visual cues were limited by a dark, moonless night. Radar information, which would have been showing on the navigator's radar scope, was not correctly interpreted. Arrival/departure charts were not studied by the pilot/copilot and were incorrectly interpreted by the navigator." The last paragraph was only one sentence long. "The crash of HAVOC 58 was caused by crew error."

Epilogue

The investigating officer discovered the obvious—HAVOC 58 flew into the mountain because of human error. Operationally, the crew made critical mistakes in flight planning, procedure verification, and crew resource management. MCI 55-130 Vol. I is very straightforward: "Final responsibility for the safe conduct of a mission rests with the aircraft commander." In assigning blame, then, the report reached the only conclusion it could. But crew error and human error are not exactly the same, although there was plenty of both to go around.

It is not known what actions the Air Force took as a result of the parallel but separate safety investigation to prevent similar accidents in the future. What is known is that they were successful in identifying other issues of real concern, factors that forced the crew to be the last link in the safety chain. Those areas included: Inappropriate pairing of inexperienced crewmembers, the frenzied pace of operations and training, disregard of the effects of fatigue when scheduling flightcrews, difficulty in accessing current and appropriate airport/aeronautical information, last-minute scheduling of missions, routinely emphasizing nonflying duties, and the "flat-area" flying mentality of the squadron. Each of these demonstrated some level of human error, and each played a part in this tragedy. More importantly, each must be remedied if accidents are to be prevented. And the prevention of accidents should be the sole reason why accidents are investigated.

References and additional reading

Shelton, Christopher. 1996. Courage under fire. *Jackson Hole Guide*. 20 August, A1-4.

Skolasky, Robert A. 1996. AFI 51-503 *USAF Aircraft Accident Investigation Board Report, C-130, SN 74-1662, Near Jackson Hole, Wyoming, 17August, 1997*. Volumes I and II, HQ 8/AF, United States Air Force.

15

Have Blue

An airshow to remember

Operator: United States Air Force
Aircraft Type: F-117A
Location: Martin State Airport, near Baltimore, Maryland
Date: September 14, 1997

A well-known sixteenth century maxim begins "...for want of a nail, a shoe was lost; for want of a shoe, a horse was lost...." and so on, leading to the eventual loss of a war and collapse of an empire. The fundamental principle, however, that seemingly insignificant oversights can have disastrous consequences, is applicable to aviation accident investigation as well as Napoleonic military history. And although widely separated in time and technology, the horse and a modern jet fighter share another basic tenet—that a vehicle can only be as strong as its weakest structural component. On a beautiful Sunday afternoon in September 1997 at an airshow at Martin State Airport near Baltimore, Maryland, twelve thousand people witnessed a spectacular reminder of these timeless doctrines.

The aircraft

The F-117A Nighthawk (see Fig. 15-1), the world's first fully operational radar-evading warplane, began life as a design study codenamed *Have Blue*. The Lockheed Advanced Development Company, better known as the "Skunk Works," in Burbank, California, built two prototypes in 1977. The use of "faceting," or shaping the structure of the airframe out of trapezoidal or triangular flat surfaces, and the liberal application of new, radar-absorbing materials and technologies made the little airplanes virtually invisible to radar.

A major drawback of its unusual shape, however, was the aircraft's inherent lack of aerodynamic stability. The severely swept-back wings incorporated two elevons on each side that controlled both roll and pitch, while a pair of all-flying rudders mounted in the shape of a "V" comprised the tail and provided lateral, or yaw, control. A quadruply redundant fly-by-wire flight control system utilized on-board computers to constantly monitor and vary control surface position, providing "artificial" stability. Without these inputs, the aircraft could never have flown, and even with this highly sophisticated system, there were some abnormal flight regimes, including high angles of attack, from which the airplane could not recover.

The *Senior Trend* program followed the successful demonstration of the two *Have Blue* aircraft, providing for five additional preproduction F-117As to be built. Minor structural and systems changes were then incorporated into each airplane as operational experience dictated. The first true production F-117A was delivered to the 4450th Tactical Group at Nellis AFB, Nevada, on September 2, 1982.

Sometimes inaccurately referred to as the Stealth Fighter, the primary role of the Nighthawk is in ground attack and suppression of an enemy's defenses. Weighing only a maximum of 52,500 pounds, the single-pilot aircraft is powered by two 10,800-pound-thrust GE F404 engines that propell it to near-supersonic speeds. Forward and downward infrared viewing systems, advanced ring-laser gyros, and Global Positioning System (GPS) receivers provide highly accurate navigational capabilities. Two internal weapons bays accommodate up to 4,000 of armament, the ordinance of choice being the incredibly precise laser-guided GBU-27 Paveway III bomb that was used so effectively against Iraq in Operation *Desert Storm*. Developed in total secrecy, a total of fifty-nine F-117As were built, the last one being delivered to the U.S. Air Force in 1990. (See Fig. 15-2.)

Flight history and background

Because of its unusual look and clandestine history, Nighthawks have enjoyed unprecedented popularity with the public as display aircraft at military airshows. By early 1997, most of the operational F-117A fleet was based at Holloman AFB in New Mexico, but two

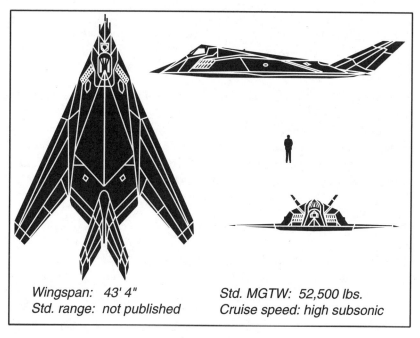

Wingspan: 43' 4"
Std. range: not published

Std. MGTW: 52,500 lbs.
Cruise speed: high subsonic

15-1 *Lockheed F-117A.*

15-2 *Lockheed F-117A in flight. . .* Courtesy IMSI

aircraft were temporarily assigned to Langley AFB in Virginia in order to more easily participate in East Coast events. One of those,
ship number 810793, was only the ninth production F-117A off
Lockheed's assembly line, and was already fourteen years old.

Major Bryan Knight, U.S. Air Force, was an F-117A pilot assigned to
the 7th Fighter Squadron and temporarily stationed at Langley AFB.
His duties there primarily involved flying demonstrations and public
displays of the Stealth aircraft. As an instructor, Maj. Knight had
flown many types of aircraft in his twelve-year Air Force career and
had accumulated almost five hundred hours flying the small, bat-
wing fighter.

The year 1997 marked the fiftieth anniversaries of the US Air Force,
Hancock International Airport in Syracuse, New York, and the 174th
Air National Guard Fighter Wing stationed there. As such, Maj.
Knight planned a routine public relations mission that included a
two-plane formation flight over Arlington National Cemetery on the
13th of September followed by a transit to and layover in Syracuse.
The next day, the local airshow appearance would be followed by
one in Baltimore, with an afternoon return to Langley AFB. On September 14, 1997, using the call sign "Stealth 71," Maj. Knight took off
from Hancock Airport in #793. By 2:30 in the afternoon, in support
of the local celebration, the F-117A's three scheduled low-altitude,
high-speed passes down Hancock Airport's main runway were complete. Maj. Knight then turned his Nighthawk south towards Martin
State Airport and the Chesapeake Air Show.

Timing is critical in producing a successful aerial display. Airborne
acts must be properly sequenced and supporting ground activities
carefully coordinated. Upon his arrival in the area, Maj. Knight informed Baltimore Approach (the controlling air traffic facility) that
he was about five minutes early for his 3:01 P.M. flyover. At first suggesting headings that would delay him briefly, the controller then requested that the F-117A fly a low pass at Baltimore International
Airport. Maj. Knight was happy to comply, but before the clearance
could be given, the organizers of the airshow were ready for his
demonstration. He was then cleared to descend to 3,000 feet and
vectored towards Martin State airport.

About six miles out, Maj. Knight visually identified the airfield, contacted the air traffic control tower, descended and began his first
flyby. Accelerating to 320 knots, he flew directly down the show
line, parallel to Runway 15, at about 1,100 feet above the ground

(agl). Very quickly reaching the airport boundary, he executed a climbing 90-degree turn to the right, followed immediately by a sharp 270-degree turn back to the left to realign his aircraft with Runway 33. The next run would be flown to the north. Slightly lower this time at 800 feet, the small jet screamed past the crowd at 350 knots, climbed rapidly in yet another series of turns and positioned itself for one final pass.

"Approach, Stealth seven one, going on my last pass now, looking for climbout from Martin State," radioed Maj. Knight. He was now accelerating in straight and level flight to 380 knots, and would very soon need to climb to his cruising altitude for the last leg of his flight home.

"Stealth seventy-one, climb and maintain one zero thousand, ten thousand feet when you're done with the low approach...I'm showing a VFR target at your two o'clock position and a mile-and-a-half, thirty-five hundred feet eastbound," replied the controller.

The F-117A was now midway down the flight line at only 700 feet above the ground, traveling at over 600 feet per second. Looking for the traffic while starting the climbout, a sudden vibration snapped the pilot's attention back inside. Four rapid "beats" shook the cockpit, and the airplane immediately rolled hard to the left. Maj. Knight was thrown forward in the seat, his body pinned down and forward by gravitational forces, restricting his vision to only the lower left portion of the instrument panel. Around him, the aircraft was shaking itself to pieces.

The left wing had violently separated about midspan, and fluttered slowly toward the ground. The severe aerodynamic pitch-up threw the main landing gear out of the wheel wells and ripped off the small nose piece of the airplane (see Figs. 15-3 and 15-4). Struggling against his restraints, Maj. Knight anxiously scanned the airspace around his aircraft, thinking he had experienced a midair collision. He realized that the aircraft was no longer moving forward, but was descending rapidly in a series of "falling leaf" stalls toward a residential neighborhood. Fighting the controls to steer the crippled fighter away from the houses and toward a nearby river, Maj. Knight made one last radio transmission. He could wait no longer.

"I just hit something! Stealth seven one's bailing out—oh God!"

The crowds of spectators first believed that the unusual flight path of the F-117A was part of the show. Vapor trailing from the fractured wing surface looked very similar to smoke trails intentionally generated by

15-3 *Stealth 71 just after the wing separates. . .* Courtesy U.S. Air Force

other aerobatic aircraft performances that afternoon. But when the canopy blew off and the pilot was catapulted out of the falling aircraft, most finally realized they were witnessing a true life-or-death drama.

As a senior engineer for the manufacturer of the Nighthawk's ejection system, one onlooker was particularly horrified by the disintegration of the aircraft right in front of him. When Maj. Knight pulled

15-4 *Stealth 71 after the gear has extended and nose separates. . .*
Courtesy U.S. Air Force

hard on handles located on either side of the cockpit, gas-fired pressure exploded inside a telescoping tube sending the pilot and the
ejection seat hurtling from the aircraft.

"He's out!" yelled the engineer, even before the parachute opened. The
ACES II (Advanced Concept Ejection Seat) system automatically determined the ejection event-time sequencing based on aircraft altitude

and airspeed. Two seconds later, pilot and seat had separated and the parachute opened. In just a few more seconds, the survival-gear package deployed and dangled fifteen feet below Maj. Knight as he drifted earthward.

The stricken airplane slammed into the community of Bowley's Quarters, exploding upon impact and sending a huge column of smoke into the sky (see Fig. 15-5). Belly flopping into a driveway, the nose of the aircraft rested in the bed of a pickup truck while one wing slid under an automobile (see Fig. 15-6). Small pieces were thrown hundreds of yards in all directions and an intense fire burned at the center of the wreckage. The house closest to the crash site immediately was engulfed in flames, and a shed and several cars were soon consumed as well. Landing just 100 yards west of the burning wreck, Maj. Knight was unharmed and was assisted away from the scene by local residents. A short time later, personnel from nearby Warfield Air National Guard Base (WANGB) escorted him back for a medical examination, and later that day he was transported to Andrews Air Force Base near Washington, D.C., for a complete medical evaluation.

15-5 *Major Knight parachuting to safety. Smoke in background is due to the ground impact of the F-117A.* . . . Courtesy U.S. Air Force

15-6 *Stealth 71 wreckage.* . . Courtesy U.S. Air Force

Incredibly, there were only a few minor injuries as a direct result of the accident. The owners of the destroyed house, having purchased it only two weeks prior, were out of town for the weekend. Most residents of the community were outside, witnessed the crash, and rushed to evacuate friends and family members from nearby homes. Rescue and firefighting personnel from the Bowley's Quarters Volunteer Fire Department and WANGB responded to the scene quickly, but in battling the blaze were exposed to intense, choking smoke. Twenty-three firefighters became ill after breathing fumes emitted from the burning jet, and two who were among the very first to arrive at the accident site were treated at Fallston General Hospital for symptoms of possible boron and radiation contamination. Fortunately, no long-term physical ailments were reported by any rescue personnel.

The investigation and findings

As discussed in the previous two chapters, military investigations must be conducted in accordance with Air Force Instructions, and differ significantly from typical civilian procedures. In this case, the Commander of the Air Combat Command of the U.S. Air Force, under the provisions of Air Force Instruction (AFI) 51-503, appointed the Accident Investigation Board. This team was comprised of investigators familiar with all aspects of military flight operations, including maintenance, aircraft structures, flight planning, and execution as well as medical factors and legal considerations. The Commander of the Twelfth Air Force appointed the Board President, Col. John Beard, and additional technical experts were assigned to the group as needed.

On scene

The two-week on-site investigation was headquartered at WANGB. Maj. Knight's description of the breakup and the extensive photographic documentation available were extremely valuable to investigators; their attention was immediately concentrated on possible structural failure of the accident aircraft.

The first order of business was to thoroughly document the location and condition of all remaining wreckage. The outboard eight-foot portion of the left wing was retrieved from a grassy area between the runway and a parallel taxiway. A 400-foot path of debris, including pieces of radar absorbing material (RAM), splintered honeycomb structure, small metal fragments, and shattered hydraulic and fuel system components led past the end of the runway, off airport property, and to the banks of adjacent Frog Mortar Creek. The canopy and ejection seat were found just on the far side of the water, with the rest of the aircraft falling nearby.

Exploding upon impact, much of the fuselage was too severely burned to allow rapid analysis of aircraft systems and structure (see Fig. 15-6). Investigative work needed to be completed quickly, however, because environmental hazards resulting from the crash needed to be cleaned up as soon as possible. Additionally, Air Force security considerations prevented anyone other than official military investigators from being near the site, so for several days many residents of Bowley's Quarters were forced out of their homes.

No prebreakup aircraft system or engine malfunctions were evident as a result of the initial wreckage examination. The investigative Board found that all flight maneuvers were conducted in accordance with Air Force and Federal Aviation Administration (FAA) guidelines and requirements, weather was not a factor, and Maj. Knight's decision to eject was timely and appropriate.

Aircraft structure

Careful analysis of videotape of the accident sequence confirmed a suspected in-flight separation of part of the left wing. A rapid oscillation, or aerodynamic flutter, of the left outboard elevon as the F-117A made its last pass down the runway led to an almost instantaneous structural failure of the wing. Normally, elevon movement and damping was controlled by a hydraulic integrated servo actuator (ISA) attached on one end to the elevon and on the other to the internal framework of the wing (see Fig. 15-7). Ascertaining the reasons for the control surface instability would prove vital to understanding the cause of this accident.

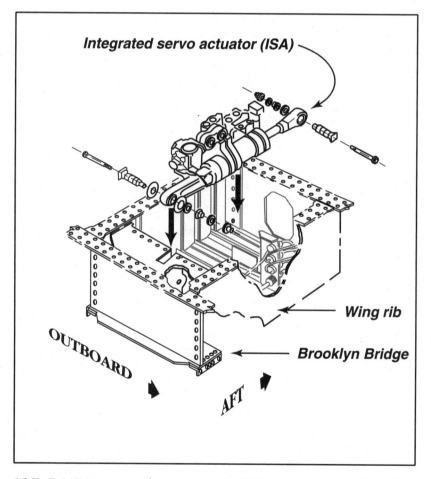

15-7 *F-117A integrated servo actuator (ISA) mounting into Brooklyn Bridge. . .* Source: U.S. Air Force Art: J. Walters

The "Brooklyn Bridge"

The Lockheed designers knew that even normal air loads imposed upon the wing in flight by the elevon, through the ISA, would be extreme. Consequently, the ISA wing attach point was designed to be extraordinarily strong and came to be known as the "Brooklyn Bridge" (see Fig. 15-8). Made of high-grade plate steel, the box structure utilized special fasteners to secure the thick upper cap and to attach the entire assembly to the adjacent wing ribs. These threaded bolts were known as "Hi-Lok" and "Taper-Lok" fasteners, and were designed to provide exceptional strength through the use of an "interference" (or extremely tight!) fit. Any hole drilled in a structure, even for a fastener, reduces its strength, and the Brooklyn Bridge assembly required many of these holes. But the interference fit virtually eliminated any void between the fastener and the structure, thus allowing the load to be centered within the hole, increasing overall load-carrying ability and the fatigue life of the bolt. Although Taper-Loks used a special taper-drilled hole and Hi-Loks did not, both types were constructed of very high strength metals, provided the necessary interference fit, and could easily be serviced by Air Force personnel.

Very early in the investigation, a careful disassembly of the left outboard ISA bay was done. Fortunately, fire damage was minimal, and all major components were easily recovered and identified. It was discovered that five of the six brackets that attach the Brooklyn Bridge to the wing were broken and the entire assembly was free to move within the bay. Additionally, of the five fasteners that secure the upper plate of the Bridge to the inboard wing attach bracket, four (all Hi-Loks) were missing (see Fig. 15-8). The lone remaining fastener, a Taper-Lok that passed through the upper cap, the upper plate of the Bridge, and the attaching bracket, had been installed but had failed after experiencing an extreme overload. The severe elongation of its hole indicated to investigators that the bolt had carried this large load for a significant amount of time prior to its failure. Later laboratory analysis indicated that the missing fasteners had been installed at one time and removed, perhaps twice.

Service history of the Brooklyn Bridge assembly

As early as 1988 problems were developing in the F-117A Brooklyn Bridge assemblies, including loose Taper-Loks and Hi-Loks and cracked support brackets. Changes were incorporated into airplanes manufactured after 1989, but early in 1995 loose Taper-Loks and cracks were found in four other early production F-117A Bridge assemblies.

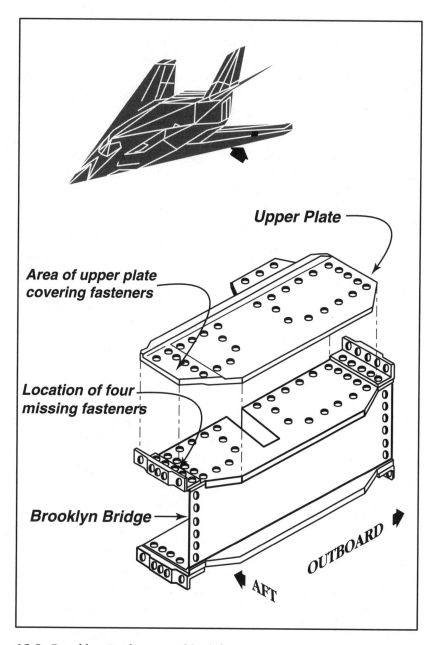

15-8 *Brooklyn Bridge assembly, left wing.. .* Source: U.S. Air Force Art: J. Walters

Aggravating the situation was the fact that the ISAs needed to be removed more frequently than planned, requiring extensive disassembly and reconstruction of the Bridge each time. Lockheed Martin proposed a structural modification in 1995 that would remedy the Brooklyn Bridge fastener problems and would also allow an ISA to be removed and serviced without removing the upper steel cap of the Bridge, but that proposal was not adopted.

During a routine inspection of the accident aircraft in January of 1996, movement was discovered in the Brooklyn Bridge assemblies of both wings, with further scrutiny revealing twenty-six loose Taper-Lok and two incorrectly installed Hi-Lok fasteners. Extensive repair to the wing boxes of aircraft #793 included a complete rebuild of both Bridges, necessitating replacement of many of the standard size fasteners with oversize ones to assure a proper interference fit. Even though inspections of the completed work were required and accomplished at many points during the process, Air Force investigators believe that it was during these maintenance procedures and the subsequent rebuild that the four missing Hi-Lok fasteners were inadvertently omitted. Those investigators noted, however, that the design and placement of the Bridge's steel upper cap made it extremely difficult to see those four particular fasteners after reassembly (see Fig. 15-8).

In mid-January 1996, an Air Force service bulletin (N-3440) was issued requiring inspection of all F-117As for Brooklyn Bridge movement and loose or inappropriate fasteners. This order required that the work be accomplished within one year and during each airplane's next depot level inspection. Aircraft #793 underwent a thorough, scheduled inspection in early August of 1996 and again later that month when scheduled for an extensive depot "configuration update." The elevon servoactuators and all attachment fittings and fasteners should have been checked for general condition, security, cracks, and loose or missing hardware each time, yet no record of any discrepancies was found. Bulletin N-3440 should have been complied with while the airplane was in the depot for other modifications, but there was no evidence of it ever accomplished.

In July of 1997, only two months before the accident, a report was made by another Air Force pilot of a potential problem with aircraft #793. While flying in formation with the F-117A, the pilot of the other aircraft noticed and reported a very unusual two-inch upward displacement of the Nighthawk's left outboard elevon during all phases of flight. A cursory inspection was done after both aircraft returned to base, but no significant maintenance actions were performed.

A review of all other maintenance records indicated no other repair work was done on the Brooklyn Bridge assemblies of the accident aircraft. The left outboard ISA that was found in the wreckage had the same serial number as the one installed during the extensive work in January of 1996, and there was no indication that the ISA compartment was even opened during the depot-level work completed in 1996.

Conclusions

Detailed wreckage analysis confirmed the sequence of events leading to the accident. The left side upper attachment fitting for the left wing Brooklyn Bridge was retained by only one fastener, leading to "softening" of the system. Air loads and limit cycle oscillations (LCO) eventually caused the fitting to partially fail, elongating the hole of the sole fastener. This in turn caused increased LCO loads. On his third pass, the attachment on Maj. Knight's aircraft failed completely resulting in a rapid increase in LCO. Almost instantly, adjacent fittings failed, rotation of the Bridge assembly began, and the oscillations of the elevon and wing coupled, leading to flutter. The violent motion of the Bridge within the wing bay caused hydraulic and electrical system failures inside the ISA bay just prior to separation of the wing.

Probable cause

In the final accident report, Col. Beard found the cause of this accident to be "...structural failure of the left wing Brooklyn Bridge assembly support brackets that attach the Brooklyn Bridge assembly to the ribs on either side of the ISA bay in the left wing due to four missing Hi-Lok fasteners." In his summary comments, he states, "It is my opinion the accident was caused by unintentional maintenance oversight during the left Brooklyn Bridge assembly reinstallation in January 1996."

As pointed out in the report, "there were three opportunities to discover that the left Brooklyn Bridge assembly was improperly installed.

"When reinstalling the steel cap in January 1996, there was the opportunity to notice that there were four extra holes...that did not correspond to holes in the steel cap. However, this would not have been readily apparent."

"An aggressive search for 'loose and missing hardware' specified on the work card during the # 4 phase inspection in July 1996 could also have detected the four empty Hi-Lok holes."

"...The abnormal upward deflection of the outboard elevon reported...in July 1997 was the result of elongation of the hole supporting the lone Taper-Lok on the upper inboard side of the
Brooklyn Bridge assembly. Had this observation been more thoroughly investigated by [Air Force] maintenance, the problem may
have been discovered."

Air Force action

Because of the suspected structural failure of the wing, the Air Force
immediately grounded all of the remaining fifty-three F-117A aircraft.
By October 4, 1997, investigators were confident that the true causes
of the accident were known, and comprehensive inspections of all
Brooklyn Bridge assemblies and attaching hardware were ordered.
No similar defects were found in any other aircraft, and the F-117As
were eventually all returned to service. Sadly, this accident was one
of five military crashes that occurred within only a four-day period.
To provide a "period of reflection" and additional safety training, Defense Secretary William Cohen ordered a "stand-down" of military
aviation, suspending all Air Force, Navy, Army, and Marine flying for
twenty-four hours.

Epilogue

It is nearly impossible to "measure" aviation safety. The best that can
be done is a comprehensive assessment, acknowledgement, and acceptance of a specific risk. In the civil sector, this process is enabled
by the free and open exchange of critical operational data and investigative reports. Industrywide cooperation usually allows effective and timely regulatory action. But each branch of the U.S. armed
forces investigates, reports, recommends, and acts upon safety initiatives solely within the confines of their own organization. The issues raised may be very relevant to civilian operators, but since they
are not the "customers" of the process, no formal mechanisms are in
place to take advantage of acquired safety information. The military
accident investigation team usually generates two reports: one for
internal use and one for release to the public if requested under the
Freedom of Information Act.

It is difficult to assess whether the lessons learned by this accident
produced any real, effective changes—the recommendations issued

by the Air Force were never publicly published. But the F-117A Nighthawk is still quietly doing its job, flying every day, continually proving its worth as an incredibly stealthy and effective weapon in the airborne arsenal.

References and additional reading

Baugher, Joe. 1999. F-117A. *Elevon: Encyclopedia of American Military Aircraft*. Available online, <http://www.csd.uwo.ca/_pettypi/elevon/baugher_us>

Beard, John H. 1997. AFI 51-503 *Accident Investigation Report, F-117A, SN 81-000793, Near Martin State Airport, Maryland, 14 September, 1997*. Volume 1. United States Air Force.

Dornheim, Michael A. 1997. Elevon vibration leads to F-117 crash. *Aviation Week & Space Technology*. 22 September, 30.

Part Four

General Aviation Accidents

16

Shattered dreams

A record-setting flight gone awry

Operator: Jessica Dubroff, her father, and flight instructor

Aircraft Type: Cessna 177B

Location: Cheyenne, Wyoming

Date: April 11, 1996

Aviators often attempt record-setting flights. They, like others, enjoy the notoriety of being the first to complete a particular flight in a certain fashion. Some succeed—others do not.

Ironically, for those who are not successful, there is sometimes greater fame in their failures than perhaps they would have achieved otherwise. While trying to be the first woman pilot to fly around the world, Amelia Earhart vanished and was presumed dead. Christa McAuliffe was to be the first school teacher in space, but lost her life in the space shuttle Challenger accident. And on a miserable stormy morning in April 1996, seven-year-old Jessica Dubroff lost her life while venturing to be the youngest person to pilot an airplane across the United States. Following the tragedy, the name Jessica Dubroff would be known in every household. America grieved over this young girl, who just days before appeared on network news programs full of life and enthusiasm as she explained her lofty goal. Her tragedy would also catch the attention of the U.S. Congress, who would react swiftly to prevent such a tragedy from ever happening again.

Flight history and background

Jessica, accompanied by her nonpilot father and her fifty-two-year-old Certified Flight Instructor (CFI), set out on this venture just one day prior to the accident. With Jessica in the Cessna 177's front left seat, the plan was for her instructor to occupy the right seat and coach her flying technique as they proceeded across the United States. Because the minimum age to hold a Private Pilot Certificate (and thus legally carry passengers) is seventeen, the CFI served as pilot in command (PIC). They departed their home-base airport, Half Moon Bay, California, at about 7:00 A.M. PDT in the CFI's Cessna 177B "Cardinal," a four-seat, single-engine aircraft (see Fig. 16-1). The rudder pedals for the left seat were fitted with three-inch aluminum extensions to accommodate the young pilot, and there were seat cushions to help Jessica see over the instrument panel.

Prior to departing on the trip, Jessica had logged some thirty-three hours of flight time with her flight instructor, a stockbroker by profession who flight instructed occasionally. The instructor's logbook showed that he had approximately 1,484 flight hours, but he had not logged any instrument time during the previous six months.

The Cessna was the loaded with promotional items, including fifteen pounds of navy-blue baseball caps which, in gold lettering stated, "Jessica Dubroff—Sea to Shining Sea." Additionally, ABC News had provided Jessica's father with a production-quality video camera and tapes to record highlights of the historic flight.

The first day's planned flying time was eight hours, with two intermediate fuel stops before arriving at that day's terminating station, Cheyenne, Wyoming. While at the second fuel stop, the airport manager observed that the CFI was "noticeably exhausted."

The three landed at the Cheyenne Airport around 5:30 P.M. MDT, where they were greeted by a large crowd of well-wishers and news media. The CFI telephoned his wife from the airport and told her that he was very tired, having awakened at 3:30 A.M. PDT that morning to appear on a national morning news program with Jessica and her father.

A Cheyenne radio station program director drove the trio to their hotel at around 7:00 in the evening. He stated that during the ride they discussed an approaching storm front, and the CFI was "very adamant" about leaving by 6:15 the next morning to beat the storm. He said that all three looked fatigued and discussed being very tired.

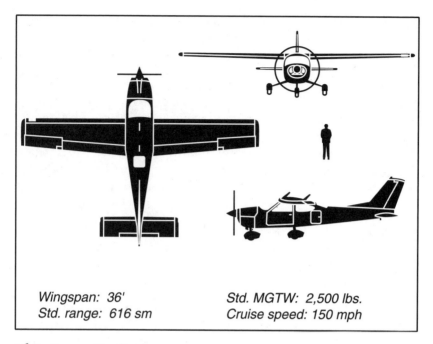

Wingspan: 36'　　　　　　Std. MGTW: 2,500 lbs.
Std. range: 616 sm　　　　Cruise speed: 150 mph

16-1 *Cessna Cardinal.*

When dawn arose, the aviators were already one hour late arriving at the airport. During a predeparture radio interview, rain streamed off the aircraft wings and formed puddles on the ramp. When the interviewer suggested that they stay in Cheyenne because of the inclement weather, Jessica's father replied that they wanted to "beat the storm" that was rapidly approaching.

The CFI contacted Casper Automated Flight Service Station just after 8:00 A.M., and stated that he was planning to depart for Lincoln, Nebraska, using Visual Flight Rules (VFR). The weather briefing included an advisory for moderate in-flight icing and moderate to severe turbulence along their intended route, and that flight precautions were in effect due to the anticipated low ceilings and/or visibilities. The weather briefer stated that were was "virtually a line of [rain showers] on a north/south line just west of your position." He stated that thunderstorms, icing, and IFR (Instrument Flight Rules) existed, and added that he was "not looking for a lot of improvement."

The weather briefer then described the current weather conditions at several points east of Cheyenne, which had fair skies. With that, the CFI replied, "Yeah, probably looks good out there from here...lookin' east looks like the sun's shining as a matter of fact." Six minutes after initiating the call, the pilot filed his VFR flight plan and hung up the telephone.

Within five minutes Jessica, her dad, and the CFI boarded the Cessna and started the engine. When the CFI called ATC for a taxi clearance, the controller asked if he had Automatic Terminal Information Service (ATIS) information "echo." The CFI replied, "negative, what is the ATIS," indicating that he was unaware of the ATIS frequency. When the controller issued the ATIS frequency, the CFI read back the wrong frequency. The controller then reissued the correct frequency.

A well-wisher videotaped N35207 as they were ready to taxi out on day two of their record-setting flight. Rain was still falling. The engine was running, but the airplane could not taxi; the nosewheel was still chocked, preventing any aircraft movement. The engine was then shut down and a lineman removed the chocks.

During taxi to Runway 30, the CFI told Cheyenne Ground Control that he was unable to tune that particular ATIS frequency. The ground controller then issued the current weather, which included winds from 280 degrees at twenty knots, occasionally gusting to thirty knots. The pilot did not acknowledge the controller's transmission, which prompted the controller to request confirmation. The CFI replied, "OK, two zero seven, are we going the right way for three zero?" The controller responded, "you are heading the right way for Runway three zero, did you get the numbers?" The pilot's response was, "We got 'em."

At 8:18 A.M. the tower controller advised the CFI that a twin Cessna had just departed Runway 30 and the pilot reported moderate low-level wind shear, which caused airspeed fluctuations of plus or minus fifteen knots. A few moments later the controller advised that the tower visibility was two and three quarter miles, and the field was IFR. Since N35207 was a VRF flight, the controller asked the pilot's intentions, whereupon the CFI replied, "OK, two zero seven would like a special IFR, um ah right downwind departure." The controller responded, "I'm not familiar with special IFR," and the pilot corrected with, "I'm sorry, special VFR."

After being issued a special VFR clearance, the pilot in command of N35207 advised Cheyenne Tower that he was ready for tak eoff. By the time the words "cleared for takeoff" were out of the controller's mouth, N35207 was already rolling down the runway.

Ground witnesses observed the airplane take off and begin a bela-bored climb and turn to the right. In general, witnesses described the Cessna as having wobbly wings, low altitude, low airspeed, and high pitch attitude, with continuous "roll oscillations" from right to left and back again. They described the weather as being quite windy with moderate to heavy mixed precipitation (rain, snow, and sleet), thunder, and lightning. As the aircraft was rolling out of the right turn at several hundred feet above ground, witnesses observed it suddenly dive nose-first towards the ground. The airplane crashed onto a residential street, about three-quarters of a mile from the de-parture end of Runway 30.

At 8:24 A.M. MDT, the dream to become the youngest pilot to fly across the country was shattered, along with the lives of Jessica Dubroff, her father, and her flight instructor.

The investigation and findings

Physical evidence at the accident scene included wreckage that was mostly confined to the immediate ground impact site. The nose sec-tion and forward cabin area were crushed and displaced rearward along airplane's longitudinal axis. Both cabin doors had crush lines indicative of a sixty-seven-degree nose-down pitch attitude at ground impact (see Fig. 16-2).

Aircraft weight and balance

The aircraft weight at takeoff was computed by investigators to be 2,596 pounds, which exceeded the Cessna Cardinal's maximum al-lowable takeoff weight of 2,500 pounds. Investigators calculated that during taxi and takeoff approximately twelve pounds of fuel were used, meaning that at the time of the accident the aircraft was about eighty-four pounds over maximum gross weight. It was determined that the aircraft was within the allowable center of gravity limits.

Weather

The accident site was approximately a mile and a half from a Doppler radar antenna. About four minutes before the crash, radar recorded moderate precipitation at the departure end of Runway 30 and very heavy precipitation in the area where the Cessna began the right turn towards the east.

16-2 *Wreckage of Jessica Dubroff's Cessna Cardinal...* Courtesy NTSB

Additionally, the 13,000-hour pilot of the twin Cessna who provided a wind shear report to ATC later told investigators that during his takeoff roll he experienced control difficulties due to strong cross-winds. After rotation, his aircraft did not accelerate as usual, and he experienced moderate turbulence and wind shear. He also noticed cloud-to-ground lightning west of the airport. Because he was aware that the Cessna 177 would soon be departing, he became concerned for their safety and provided a pilot report (PIREP) to ATC. Another pilot, the captain of a United Express Beech 1900, landed about 8:20 A.M. After seeing lightning within a mile and a half of the airport and hearing the PIREP, he decided to delay his departure. From these observations, the NTSB concluded that the accident sequence took place near the edge of a thunderstorm.

Human factors issues

Human factors were investigated to better understand what might have influenced the pilot to take off into an approaching thunderstorm. The NTSB looked at fatigue, the effects of time pressures from an overly ambitious itinerary and media attention, and aeronautical decision making.

Fatigue

The NTSB felt that the CFI's sleeping schedule in the days prior to accident flight may have led to fatigue. He received only about six hours of sleep each night in the three days prior to the start of the trip on April 10. On that morning he arose at 3:30 A.M., and later in the day told several people that he was tired. The NTSB stated that since people tend to underestimate their level of tiredness, the CFI's mentioning to several people that he was "really tired" likely reflected a high level of fatigue the day before the accident.

On the morning of the accident, the CFI checked out of the hotel at 6:22 A.M. The NTSB conceded that the type and quality of his sleep during the Cheyenne layover was unknown, but they also noted that the types of mistakes the CFI made that morning are consistent with a lack of alertness. The NTSB enumerated these errors: starting the airplane with the nosewheel still chocked; requesting a taxi clearance without having obtained the ATIS; reading back a radio frequency incorrectly; accepting a radio frequency that he could not dial on his radio; failing to acknowledge, as requested, the weather information provided by the controller; asking "are we going the right way;" failing to stop at the end of the runway; and using incorrect phraseology when requesting a "special IFR" clearance.

The NTSB stated that the number and variety of these errors were consistent with performance degradation produced by fatigue. However, they also stated that other explanations for this performance could include reasons such as "rushing, distraction from tasks, or the influence of habitual bad flying practices." In explaining this last item, investigators learned that the CFI had executed unpublished instrument approaches when weather was below VFR minimums and he once attempted to taxi with a tow bar still attached to the airplane. The Board also discovered that just one week before the accident, while making a flight with several reporters, he neglected to perform an engine runup prior to departure and forgot to close the airplane door before takeoff.

In their final analysis, however, the NTSB felt there was insufficient evidence to conclude that fatigue was a factor in the accident.

Itinerary and media attention

As shown in Fig. 16-3, the planned itinerary for this record-setting flight called for approximately fifty-one hours of flying time over

Date	Origination	Destination	Intermediate Fuel Stops	Flight Time
4/10/96	Half Moon Bay CA	Cheyenne WY	Elko NV Rock Springs WY	8:00
4/11/96	Cheyenne WY	Fort Wayne IN	Lincoln NE Peoria IL	7:30
4/12/96	Fort Wayne IN	Falmouth MA	Cleveland OH Williamsport PA	6:00
4/13/96	Falmouth MA	Clinton MD	Frederick MD	3:00
4/14/96	Clinton MD	Lakeland FL	Raleigh NC Charleston SC Jacksonville FL	6:45
4/15/96	Lakeland FL	Houston TX	Marianna FL Mobile AL	7:00
4/16/96	Houston TX	Sedona AZ	San Angelo TX Albuquerque NM	8:00
4/17/96	Sedona AZ	Half Moon Bay CA	Lancaster CA	5:00

16-3 *Itinerary of Jessica Dubroff's transcontinental flight...* Source: NTSB

eight days, with no planned days off. Daily scheduled flight times ranged from three to eight hours.

Documents found in the wreckage included a handwritten letter from Jessica to the President of the United States, which stated, in part, "May I visit you at the White House and even more so, will you fly with me for simply fifteen to twenty minutes...? I am scheduled to arrive in D.C. on Saturday April 13, 1996." Additionally, their schedule called for arriving at the Lakeland Florida Airport by 1:00 P.M. on Sunday April 13, when the airport closed for the beginning of the annual "Sun 'n Fun" fly-in.

Media attention surrounding the trip had grown, and the ABC News camera loaned to Jessica's father also served as a constant reminder of the news value of this flight. Additionally, postaccident examination of his personal effects revealed "numerous slips of paper with appointment times and dates of TV interviews" in the pocket of the shirt he was wearing. One was with a major television network, and was to be conducted in Fort Wayne the evening of the accident. Another listed a "big TV special" that was to be taped in Massachusetts the following night. The NTSB noted that the widespread media attention could well have affected the decision-making ability of the pilot in command by increasing the perceived importance of maintaining the ambitious schedule. "The Safety Board concludes that the itinerary was overly ambitious, and that a desire to adhere to it may have contributed to the pilot in command's decision to take off under the questionable conditions at Cheyenne."

Aeronautical decision making

The NTSB observed that training pilots in specific Aeronautical Decision Making (ADM) skills has proved quite effective for pilots of emergency medical helicopter and air carriers operations. Although the FAA has published an Advisory Circular (AC) pertaining to ADM for general aviation pilots, the NTSB was concerned that a number of such pilots have not been exposed to this training. "This accident demonstrates the need for continued efforts in the area of aeronautical decision making for general aviation pilots," said the NTSB. "The circumstances of this accident could be instructive to other general aviation pilots in raising their awareness of potential decision making errors."

Conclusions

The investigation found no evidence that aircraft maintenance was a factor. Because the ground temperature was above freezing, and because of the short duration of the flight, it was concluded that airframe icing was not a factor. Autopsy findings of the CFI revealed injuries that were consistent with his handling the flight controls at time of impact. Such injury patterns were not found on the seven-year-old pilot trainee.

Witness statements indicated that the aircraft climb rate and airspeed were slow, and once it turned to an easterly heading, it rapidly rolled off on a wing and descended steeply to the ground in a near vertical flight path. Maneuvers such as these are consistent with an aerodynamic stall, so the Safety Board investigated those factors that could have led to loss of lift.

From analysis of eyewitness statements, PIREPs, and radar data, the Safety Board concluded that the accident airplane most likely encountered light to moderate rain as it began its takeoff roll, and encountered increasing amounts of precipitation at the time of the accident. Ground-based radar reflectivity returns indicate that just before the accident the airplane experienced a precipitation rate equivalent to 3.146 inches per hour. The NTSB cited NASA research that shows that such precipitation rates can reduce lift by as much as three percent, and increase stall speed by about one and a half percent.

The NTSB estimated that the aircraft was being turned at roughly a twenty-degree bank angle as the pilot made the right turn to the east. The NTSB computed that this bank angle would have increased the Cessna's stall speed about three mph, from about fifty-nine mph to about sixty-two mph. The Safety Board also determined that because the aircraft was over its maximum allowable gross weight, the stall speed would have been increased by another two percent. Collectively, the effects of rain, bank angle, and excess weight would have increased the stall speed to about sixty-four mph. The combination of the Cheyenne Airport's high elevation (6,156 feet msl) and a temperature of forty degrees Fahrenheit created a density altitude of 6,670 feet. This high-altitude condition would affect climb performance by reducing the rate of climb to roughly 387 feet per minute (fpm) after takeoff, compared to 685 fpm at sea level.

The Safety Board acknowledged that the airplane still should have been able to climb and turn in spite of those factors that increased stall speed and decreased climb performance. Their analysis focused further on possible reduction in engine performance due to carburetor icing and/or overrich fuel/air mixture, and the effects of reduced visibility, fluctuating winds, and the pilot's lack of experience in takeoffs from high-density airports.

Reduced engine performance

When the Cessna departed Cheyenne, environmental conditions were present which were conducive for formation of carburetor icing (temperature forty degrees, dew point thirty-two degrees, and moisture in the air). Thus, carburetor icing may have formed which would have reduced the available power at takeoff. Examination of the wreckage found the carburetor heat valve in the "off" position. "The pilot's failure to stop at the end of the runway before his takeoff roll suggests that he did not perform a pretakeoff checklist, which would have included a magneto and carburetor heat check (turning on the carburetor heat)," stated the Safety Board. They stated, however, that they could not conclusively determine its position during the takeoff sequence.

Carburetors for reciprocating engines, such as the one in N35207, are typically calibrated for sea level pressure to meter the correct fuel/air mixture when the mixture control is in the "full rich" position. The NTSB stated that when taking off from high-altitude airports such as Cheyenne, failure to lean the mixture could result in an appreciable loss of power and reduced climb performance due to an overly rich mixture. Investigators found the mixture control in the full rich position at the accident site, but noted that it could have been properly set prior to the accident, but upon ground impact moved to the forward (full rich) position. They added, however, "The pilot's failure to stop at the end of the runway before his takeoff roll, which would have been the most common and appropriate time to adjust the fuel/air mixture, further suggests that he did not properly lean the mixture."

The Safety Board concluded that either carburetor icing or an overly rich mixture could have reduced the engine power sufficiently to cause a loss of climb capability. They acknowledged, however, that because the settings of these knobs and valves during takeoff they could not conclusively determine whether engine performance was actually affected by carburetor icing or an overly rich fuel mixture.

Operational factors

The NTSB concluded the natural horizon was not visible because precipitation degraded in-flight visibility. However, they felt that the pilot in command could have maintained visual ground reference by looking out the side window. They stated that this cross-scanning could have been disorienting to the pilot "because of his need to scan to his left to see flight instruments in front of the pilot trainee and [then scan] to his right to see the ground, as he attempted to operate the airplane at low speed, with a lower-than-normal climb rate, with the distractions of the rain and ice impacting the airplane, and in instrument meteorological conditions."

The crosswind component was twenty-one to twenty-three knots at the time of N35207's takeoff. Post-accident analysis showed that these winds were about five knots stronger than when the twin

Cessna departed minutes earlier and encountered strong crosswinds, wind shear, and turbulence. The Safety Board surmised that N35207 would have also encountered these adverse conditions. They concluded that the gusty wind conditions, turbulence, and wind shear would have made it difficult to maintain a constant airspeed and rate of climb, which could have resulted in an unintended reduction in airspeed to below the airplane's stall speed.

The wind conditions could have affected the CFI's perception of airspeed as well, according to the Safety Board. "What was initially a crosswind during the takeoff roll and initial airborne phase became a tailwind after the airplane began its right turn." They concluded that because the pilot was most likely looking outside during his VFR departure, he may not have been adequately monitoring the airspeed indicator, or may have had difficulty in doing so due to the turbulence. The Board felt that this reliance on outside cues may have mislead the CFI, because he may have misperceived the increased ground speed (due to tailwind) as an increased airspeed. "Accordingly, the Safety Board concludes that the right turn into a tailwind may have caused the pilot in command to misjudge the margin of safety above the airplane's stall speed. In addition, the pilot may have increased the airplane's pitch angle to compensate for the perceived decreased climb rate, especially if the pilot misperceived the apparent ground speed for airspeed, or if the pilot became disoriented."

The investigators also noted that the CFI's lack of experience at high-altitude airports could have affected his actions. The high density altitude would have significantly reduced the airplane's climb rate compared to the low-elevation airport where the CFI and aircraft were based. The NTSB stated, "This reduced rate of climb might well prompt a person who was inexperienced with high density altitude takeoffs to raise the nose of the airplane in an attempt to increase the rate of climb, thereby further decreasing the airspeed. Therefore, the Safety Board concludes that the high density altitude and possibly the pilot in command's limited experience with this type of takeoff contributed to the loss of airspeed that led to the stall."

The NTSB was unable to determine which of the above factors, or combination of factors, resulted in reduction in climb speed to below the stall speed. But in the final analysis, they did not hesitate to state the accident was caused by "the pilot in command's failure to ensure that the airplane maintained sufficient airspeed during the initial climb and subsequent downwind turn to ensure an adequate margin above the airplane's stall speed, resulting in a stall and collision with the terrain." They also stated that the pilot in command "inappropriately decided to take off under conditions that were too challenging for the pilot trainee and, apparently, even for him to handle safely."

Conclusions and probable cause

The NTSB determined the probable cause of this accident was the pilot in command's improper decision to take off into deteriorating weather conditions (including turbulence, gusty winds, and an advancing thunderstorm and associated precipitation) when the airplane was overweight and when the density altitude was higher than he was accustomed to, resulting in a stall caused by failure to maintain airspeed. Contributing to the pilot in command's decision to take off was a desire to adhere to an overly ambitious itinerary, in part, because of media constraints.

Recommendations

From their investigation of this accident, the NTSB recommended that the Aircraft Owners and Pilots Association (AOPA) and the National Association of Flight Instructors disseminate information about the circumstances of this accident and continue to emphasize to their members the importance of aeronautical decision making.

It was recommended that the FAA expand the development and increase the dissemination of educational materials of the hazards of fatigue to the general aviation piloting community. Also, the FAA should incorporate lessons learned from this accident into educational materials on aeronautical decision making (A-97-20 and 21).

Industry action

In conducting aviation safety meetings and seminars, the FAA now uses several new videos to educate general aviation pilots on the effects of fatigue and other aeromedical issues that affect the personal decision to fly. The Aviation Safety Program's "Back to Basics III" education effort focuses on Aeronautical Decision Making, especially regarding takeoffs and landings, fuel management, and developing personal operational limits for the typical private pilot.

Epilogue

In response to this accident, in October 1996, the United States Congress passed HR3276, The Child Pilot Safety Act, which prohibits "record setting flights" by unlicensed pilots (including underage student pilots). It also required the FAA to conduct a study on the overall implications of children flying aircraft, which had not yet been published as of the fall of 1999.

References and additional reading

Asker, James R. 1996. Emotional issue. *Aviation Week & Space Technology*. 6 May, 19.

Federal Aviation Administration. 1998. *NTSB Recommendations to FAA and FAA Responses Report.* <http://nasdac.faa.gov/ >

Garrison, Peter. 1997. Fifteen minutes. *Flying*. August, 109-111.

Howe, Rob, and Stambler, Lyndon. 1996. Final adventure. *People*. 29 April, 88.

National Transportation Safety Board. 1997. *Aircraft accident report: Inflight loss of control and subsequent collision with terrain. Cessna 177B, N35207. Cheyenne, Wyoming. April 11, 1996.* NTSB/AAR-97/02. Washington, D.C.: NTSB.

National Transportation Safety Board. 1997. Public docket. *Cessna 177B, N35207. Cheyenne, Wyoming. April 11, 1996. SEA96MA079.* Washington, D.C.: NTSB.

Phillips, Edward H. 1996. Girl's aircraft overweight. *Aviation Week & Space Technology*. 22 April, 28.

——. 1997. Instructor faulted in Dubroff crash. *Aviation Week & Space Technology*. 10 March, 36.

17

Falling star

John Denver's final flight

Operator: John Denver

Aircraft Type: Long-EZ

Location: Pacific Ocean near Pacific Grove, California

Date: October 12, 1997

During the 1970s and 1980s, John Denver's music inspired his millions of fans worldwide. Songs emphasizing his love of nature and concern for the environment were as popular as the more traditional folk music that started his rise to fame. In 1967, he wrote the number-one hit "Leaving on a Jet Plane" that was recorded by "Peter, Paul, and Mary." Singer Mary Travers of that trio would later say of Denver, "He was the Jimmy Stewart of folk music."

Like Jimmy Stewart, John Denver had a love for airplanes. "Over the years he collected vintage biplanes, two Cessna 210s, a Christen Eagle aerobatics plane, and, fatefully, the tiny Long-EZ plane in which he died," wrote *People* magazine.

Indeed, it was a sad day in the autumn of 1997 when the world learned that another star had fallen. And while millions would ask the proverbial question "why," the NTSB's findings would prove to be amazingly straightforward. And considering Denver's flying experience and his love and respect for aviation, the findings would prove to be equally as disturbing.

Flight history and background

It was just after noon on a breezy fall Saturday when singer John Denver arrived in his Lear Jet at the Santa Maria, California, Airport. He eagerly anticipated taking delivery of his new amateur-built Long-EZ aircraft purchased only two weeks before, now sporting a custom paint scheme and a specially requested registration number, N555JD.

Elated with the new paint job, Denver treated all those who worked on the aircraft to a pleasant lunch. Around two o'clock the group returned to the Santa Maria airport where Denver and another pilot began a pre-flight inspection of N555JD. Before Denver could fly the aircraft home, he would first undergo a checkout by the other pilot to make sure he was proficient with the cockpit layout and flight characteristics.

With the singer seated in the cockpit, the two began methodically re-viewing each instrument and aircraft system. An extra seat cushion was placed behind Denver's back to help him reach the Long-EZ's rudder pedals.

By three o'clock they had started the Long-EZ's Lycoming engine and taxied for departure. The engine checked out well during the pretakeoff engine "runup" inspection. "John's takeoff was very good," later recalled the checkout pilot. "We did some 360-degree turns both right and left. We did some slow flight. John seemed very comfortable with the plane." In spite of a gusty thirty-mile-per-hour wind, Denver's first landing was good. Their next takeoff and land-ing also went smoothly. "The landing was very good and I asked John how he felt about it," said the checkout pilot. "He said he was comfortable with the plane and didn't think he would have a prob-lem flying it. I felt good with his flying also." According to an air traf-fic control (ATC) tower tape recording, the checkout flight totaled about ten minutes. (See Fig. 17-1.)

They returned to the fixed base operator (FBO) where the checkout pilot deplaned. Before Denver departed for his home base, Monter-rey Peninsula Airport, California, the checkout pilot told him that the aircraft had ten gallons of fuel (one hour's endurance) in the right tank, and five gallons (thirty minutes' endurance) in the left tank. The pilot advised Denver that he should think of the five-gallon tank as strictly "reserve fuel," so Denver switched to the right tank.

Before Denver departed for Monterrey, however, the checkout pilot was haunted by Denver's difficulty in reaching the fuel selector han-

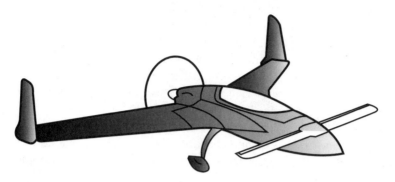

17-1 *Long-EZ. . .* Art: J. Walters

dle. It was located behind the pilot's left shoulder and Denver's use of an extra seat cushion made it even harder to reach. In addition to being out of the pilot's view, the handle was stiff to operate. Although confident of Denver's ability to fly the craft, this concern prompted the checkout pilot to ask Denver to phone him once he safely arrived at Monterrey. That evening the checkout pilot was relieved when the famous singer called to say that the flight home went well, and that he was really pleased with his new Long-EZ.

The next morning Denver arose and reportedly said, "I'm going to have a great Sunday. I'm going to play golf, and then I'm gonna fly my new bird." And indeed, after playing eighteen holes of golf, Denver drove his silver convertible Porsche to the Monterrey airport.

An aircraft mechanic helped Denver pull N555JD out of the hangar, and the two chatted as Denver spent about twenty minutes conducting a preflight inspection. They discussed the inaccessibility of the fuel selector handle and its resistance to being turned. Denver told the mechanic that when he was away on a performing tour in a few weeks, he would have the selector handle moved to a more accessible location in the cockpit. For the time being, though, Denver had an idea: he would try vice grip pliers to see if they could help him extend his reach and improve his leverage on the handle. When that did not solve the problem, he told the mechanic that he would use the autopilot in flight, if necessary, to hold the airplane level while he craned around and switched fuel tanks.

Denver borrowed a fuel sump cup from the mechanic and drained a fuel sample to check for contamination. He then climbed into the front cockpit seat, got securely buckled in, and continued his preflight duties. When the mechanic remarked that the fuel sight gauges were on the sidewalls of the aft cockpit, Denver responded that he could not see the gauges from his position, and asked the mechanic to read the quantities of each tank. The fuel gauges had no markings to indicate how they were calibrated, so the mechanic estimated that the right tank had less than a half tank of fuel, and the left tank contained less than a quarter tank. The mechanic asked Denver if he would like to add fuel, but Denver declined, saying that he would only be flying for about one hour. As a final gesture, the mechanic provided Denver with his inspection mirror so that he could look over his shoulder to see the fuel sight gauges.

The canopy was closed and Denver started the Lycoming engine. As the mechanic walked to close the hangar doors, he heard the engine quit after running for about ten seconds. Watching from a distance, he noticed the pilot had turned toward the fuel selector handle. In a few seconds, the engine had been restarted and Denver signaled "okay" to the mechanic. The mechanic watched as Denver taxied for departure.

"Tower, this is long easy triple five juliet delta, ready for takeoff, two eight left. Like to stay in the pattern, do some landings—touch and goes."

At twelve minutes after five, Monterrey tower cleared N555JD for takeoff. About fourteen minutes later, after his third touch and go, Denver stated, "Tower, juliet delta would like to continue straight out and take a flight around the point, if I may."

"Long easy five juliet delta, roger, squawk zero three six seven," replied the tower. John Denver acknowledged that transmission, but about forty-five seconds later the controller was still not receiving N555JD's transponder return. "Long easy five juliet delta, I'm not receiving your transponder."

Denver likely then turned the transponder on and asked, "How about now, sir?"

"Long easy five juliet delta, thank you. I have it now."

"Long easy five juliet delta, contact departure on one two seven point five," stated the controller. There was no reply.

Thirty seconds later, the controller stated, "Long easy five juliet delta, contact Monterrey approach on one two seven point five." This transmission also went unanswered, and the controller noticed that the primary radar return for N555JD had been lost. He notified his supervisor of the situation.

Around this time, several people walking along the rocky coastline of the Pacific Ocean heard the distinctly loud sound of the Long-EZ's engine and looked up to spot it. Most observers' estimates placed the aircraft between 350 to 500 feet above the water. Suddenly, at least eight people heard a popping noise or engine backfire, followed by an immediate reduction in the engine noise level. As the airplane came within about 100 yards of the shoreline, it was reported to have banked to the right, then pitched up, followed by a steep nose-down descent towards the water. "We thought the pilot was doing some low altitude maneuvers or stunt flying and that he would pull out of his downward descent," noted one bystander. "He did not. He crashed about 150 yards from the shore in about thirty feet of water. There was no explosion, just a tremendous splash."

For several minutes after the impact, one witness described that "the water still showed signs of upwelling." Floating on the water's surface were pieces of Styrofoam, fiberglass, and a few scattered bird feathers. The time was about 5:28 in the afternoon.

The investigation and findings

The NTSB's Los Angeles Regional Field Office responded to the accident. Divers from the National Oceanic and Atmospheric Administration, Monterey Bay National Marine Sanctuary, used an underwater video camera to document the wreckage. One of the divers stated that there were large rocks located only about ten feet below the surface. All major structural components of the airframe were found in a highly fragmented state near the engine, which had separated from the airframe. The landing gear assembly was separated from the fuselage, and the right wheel and brake were separated from the gear leg. The nose gear was found in the retracted position.

The shattered wreckage was then recovered from the rocky ocean floor and taken to an empty hangar at the Monterrey Airport (see Fig. 17-2). "There were very few intact parts," stated an investigator as he described what he found. "Most of the pieces consisted of shattered fiberglass skin, and broken bits of foam which, once identified, could be laid out on the hangar floor, and could then be easily seen to be, for example, a left wing, or whatever." Each piece of the flight control systems was identified and determined to have failed from the high impact forces. There was no evidence of any preexisting cracks or flaws in the airframe. The engine was examined and no discrepancies were found. Damage to the propeller was indicative of very little or no power to the prop at the time of impact.

Very little of the forward fuselage was recognizable. The investigator who did the wreckage reconstruction speculated that this extensive damage occurred when the aircraft impacted the submerged rocks, and the rear-mounted engine "crashed through center-section spar, the back seat bulkhead, the front seat bulkhead, the pilot, the instrument panel, and the nose area of the plane."

The aircraft

"Amateur-built" aircraft are typically built by aviation enthusiasts who possess the strong desire to build their own aircraft. The hobbyist usually purchases plans or a kit from a designer and then spends countless hours over the next several years constructing the aircraft. Amateur-built aircraft have gained tremendous popularity in recent years, with the Experimental Aircraft Association (EAA) esti-

17-2 *Reconstructing Long-EZ N555JD. . .* Source: NTSB

mating that about 95,000 plans or kits have been sold to prospective amateur builders in the last decade alone. Overall, there have been about 4,500 sets of Long-EZ plans sold, and about 1,200 Long-EZ aircraft have been constructed.

The Long-EZ was designed by Rutan Aircraft Factory as a single-engine, two-seat tandem cockpit aircraft. Instead of a conventional horizontal tail surface, the Long-EZ utilizes a "canard" wing, which is a lifting airfoil mounted to the front fuselage of the aircraft. The airframe is constructed of composite materials, including shaped foam covered with fiberglass/epoxy resin. The swept-wings are laminar flow airfoils, which are often used in high-performance aircraft. They are designed to minimize drag but they are also sensitive to boundary layer separation, or in nonprofessionals' terms, they can easily be stalled. The wingtips are twisted upwards to form winglets, and at the trailing edge of each winglet is a rudder. The Long-EZ's rudder system, controlled by conventional rudder pedals, is very sensitive in low-speed flight. Pitch and roll are controlled by a flight control side stick, located on the right side of the forward and aft cockpits.

The conception of John Denver's Long-EZ, serial number 54 was in Spring of 1980, when amateur builder Adrian Davis ordered the plans from Rutan. He then purchased several thousand dollars worth of necessary parts and supplies from an aircraft supply company in Illinois. The parts were then trucked to Mr. Davis, who at the time, resided in Memphis, Tennessee. In June of that year, the FAA assigned an aircraft registration number to the yet-to-be-built airframe, N5LE.

For the next seven years, the aircraft slowly came together. In June 1987, the FAA issued a Special Airworthiness Certificate for Mr. Davis' airplane in the "experimental" category. The experimental designation is typically issued for amateur-built aircraft, as they are not required to meet the stricter certification standards of factory manufactured aircraft. Since the FAA issues experimental airworthiness certificates for amateur-built aircraft with the original builder's name as the manufacturer, John Denver's aircraft was officially listed as an "Adrian Davis Long-EZ."

In March 1994, the Long-EZ was sold to a California veterinarian who changed the registration number to N228VS. On September 26, 1997, the veterinarian sold the aircraft to John Denver, and the aircraft was renumbered as N555JD.

The aircraft was equipped with a Lycoming O-320-E3D engine that produced 150 horsepower (hp). However, the owner's manual specified that the aircraft was designed for either a Continental O-200 engine with 100 hp, or the Lycoming O-235 engine with 115 hp. The airplane's designer wrote the NTSB and reaffirmed that only two engines were approved for the aircraft; however, he added that "...there are substantial margins in the design" of the aircraft to accommodate the larger engine.

Fuel selector handle and valve assembly

The aircraft was constructed with another unauthorized modification. The fuel selector handle was located on the bulkhead behind the pilot's left shoulder, contrary to designer's drawings which specified that it be located between the forward cockpit pilot's legs (see Fig. 17-3). The fuel selector handle was connected to the actual fuel selector valve via a forty-five inch steel and aluminum tubing and universal joint.

Investigators talked with the veterinarian who sold the airplane to Denver and learned the aircraft's original owner/builder placed the fuel handle in that location because he did not want fuel lines in the cockpit area. The veterinarian told investigators that his personal technique for switching tanks in flight was to engage the autopilot to free his right hand, then reach behind his left shoulder and grab the selector handle. He stated that the handle was "firm to turn with good detents."

Other pilots who had flown this aircraft reiterated to investigators that in order to switch fuel tanks in flight, a pilot had to:

1. Remove his hand from the right side control stick if he was hand flying the aircraft.
2. Release the shoulder harnesses.
3. Turn his upper body ninety degrees to the left to reach the handle.
4. Turn the handle to another position.

The pilot who checked-out Denver the day before the accident told investigators that he was not pleased with the handle's location, and because of difficulties reaching the selector, he never used it in flight. Compounding the problems of the handle being hard to reach and turn, the fuel selector handle in N555JD was not placarded or marked for any of its operating positions. Investigators learned that this had resulted in problems for at least two other pilots. In one

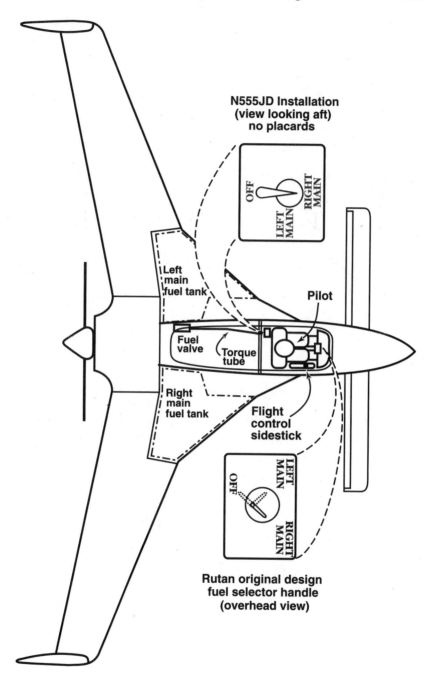

**N555JD Installation
(view looking aft)
no placards**

OFF / LEFT MAIN / RIGHT MAIN

Left
main
fuel tank

Pilot

Fuel
valve

Torque
tube

Right
main
fuel tank

Flight
control
sidestick

LEFT MAIN / OFF / RIGHT MAIN

**Rutan original design
fuel selector handle
(overhead view)**

17-3 *Diagram of N555JD's (nonstandard) and standard installations of fuel tank selector.* . . Source: NTSB Art: J. Walters

case, because of the lack of placards, the pilot inadvertently selected a tank that contained only minimum fuel; he subsequently ran that tank dry in flight and almost crashed before reselecting another tank. Investigators noted that the aircraft was supposed to have "a placard associated with the Rutan design [which] clearly identifies the fuel selector handle positions." Further adding to the confusion was that the fuel sight gauges, which were only visible from the aft cockpit, were not inscribed with calibration marks.

The fuel selector valve was recovered and sent to the NTSB materials lab in Washington, D.C. The valve was found to be frozen or "locked" in an intermediate position and could not be moved. The selector valve port from the right fuel tank supply was about thirty-three percent open to the engine feed line and the left tank port was about four percent open. The valve was taken to an engine test cell and it was shown that full power could be attained on the test engine as long as there was fuel in both tanks. However, with fuel in only one tank and the other fuel line simulating an empty tank, the engine lost power.

Pilot information

To the world, he was known as John Denver, but his legal name was Henry John Deutschendorf, Jr. His father, a retired U.S. Air Force test pilot, taught him to fly in 1976. He held a private pilot certificate with an instrument rating, along with credentials to pilot sea planes, gliders, and multiengine aircraft, including his own Lear Jet. Although he had about 2,800 flight hours, before the accident flight he only had a little more than one hour of piloting experience in a Long-EZ.

In 1993 and again in 1994, Denver was caught by authorities while driving his car under the influence of alcohol, prompting the FAA to revoke his FAA medical certificate. After being convinced of his rehabilitation, however, the FAA reissued his medical certificate in October 1995. In a letter to Denver, the FAA reminded him that "Continued airman medical certification remains contingent upon your total abstinence from use of alcohol." When that certificate expired several months later, Denver underwent another FAA medical exam in June 1996, and was reissued a third class FAA medical certificate. However, in November 1996 the FAA Civil Aeromedical Division sent him a letter by certified mail, explaining that they had received an interim report from Denver's physician, as required by his conditional recertification. The physician's letter, however, informed the FAA that while there was "no abuse" of alcohol, the singer "in general, averages two to four drinks of wine or beer/week when he's traveling." Based on that report, the FAA's letter told Denver that "we have no alternative except to determine that you do not meet the medical standards" set forth in the Federal Aviation Regulations. That letter was never claimed, and three weeks later it was returned to the FAA.

In March 1997, the FAA again sent Denver a similar letter, which cautioned that "...in view of this finding of disqualification, the exercise of the privileges of your certificate would constitute a violation [of the FARs.]" The return receipt for the certified letter was examined by Safety Board investigators; however, the signature of the person who signed for the mail was illegible.

There were no indications that alcohol played a role in the accident. In an earlier letter to the FAA, a close friend of Denver wrote to support Denver's sobriety. He stated, "Having traveled nearly full-time with John, I know that never were alcohol or drugs even close to the operation of flying his plane; John has always been fastidious about this." Significantly, post-mortem toxicological analysis was negative for all screened drugs and ethanol.

Fuel consumption tests

The pilot who checked out Denver the day before the accident indicated that the Long-EZ had fifteen gallons of fuel when Denver departed Santa Maria for Monterrey. Considering the last known refueling and the fuel burn on flights since that time, the NTSB later more precisely estimated that the aircraft had about twelve and a half gallons in the right tank and three and a half gallons in the left tank when it departed Santa Maria. They calculated that during the flight to Monterrey the Long-EZ consumed about six gallons. The checkout pilot advised Denver to use fuel from the right tank, since it contained the most fuel, and he observed Denver switch to that tank before engine start.

To compute the estimated fuel burn for the accident flight, an engineering representative of Scaled Composites, Inc. (Rutan Aircraft) flew the same profile (engine start, taxi, engine runup, takeoff and three touch and goes, and a pattern departure) in a Lycoming O-320-equipped Long-EZ. The fuel burn was about three and a half gallons. He also noted that when a fuel tank was intentionally run dry, the time between selecting a tank with fuel and the resumption of engine power was about eight seconds.

Other tests and research

The NTSB obtained an audio recording of the ATC communications between John Denver and the Monterrey Tower. In an attempt to learn about any potential powerplant abnormalities, the NTSB conducted a sound spectrum analysis of the engine and propeller. The first six radio transmissions examined were when Denver was in the traffic pattern doing touch and go landings, and the three remaining transmissions were made as he was departing the local traffic pattern. Their analysis showed the engine RPM was fairly constant throughout the fourteen minutes sampled, ranging from 2,100 to 2,280 RPMs.

Because bird feathers were found commingled with the wreckage, the Safety Board examined the wreckage for evidence of a possible bird strike. However, the canopy and the leading edges of the canard and wing sections were destroyed, effectively obliterating any opportunity to find physical evidence of a bird strike. The curator of the local Museum of Natural History was asked to examine the feathers. Interestingly, the cushion that Denver had propped behind his back was discovered in the debris, but had been torn open. According to the cushion material tag, it was filled with goose feathers; however, the curator also found duck feathers in the cushion. The cushion feathers matched those that were found around the wreckage.

Conclusions

The NTSB concluded their investigation by focusing on two main areas: fuel starvation and loss of aircraft control.

It was a simple math problem to determine fuel on board when Denver departed Monterrey. Based on the aircraft's fueling history, the mechanic's observation of the fuel levels at engine start, and the normal fuel consumption rates of the airplane's engine, the NTSB concluded that the aircraft likely had about three and a half gallons of fuel available from the left tank and about six and a half gallons of fuel in the right tank prior to departure. Using the test data provided by the engineering pilot from Scaled Composites, Inc., the fuel burn for the accident flight was around three and a half gallons. It was further noted that when the fuel selector valve was placed in an engine test cell, a test engine lost power when one tank was simulated to run dry.

The Safety Board also considered the statement of the mechanic who accompanied John Denver as he did the preflight inspection. When Denver asked him how much fuel was on board, the mechanic replied that there was less than a quarter of a tank in the left tank and less than half a tank in the right tank. The Safety Board stated, "The technician estimated the fuel quantity based on the assumption that the presentations on the unmarked sight gauges were linear. However, the Long-EZ fuel tank sight gauges are not linear, and examination of other Long-EZ sight gauges revealed that the actual fuel on-board the airplane would have been much lower than the technician's estimate." (See Fig. 17-4.)

The mechanic told investigators that he and Denver had fiddled with the fuel selector handle, and when he walked away from the plane, although the handle's operating positions were unmarked, he thought it was in the "off" position. When the engine stopped shortly after Denver started it, the mechanic began walking back towards the aircraft to suggest that the fuel handle may be in the "off" position. However, before he could reach the aircraft, the pilot had already made another handle selection and restarted the engine. When rotating the fuel selector handle, the left tank selection was the detent immediately next to the "off" position. It is speculated that when the engine quit, Denver simply turned the handle to the first available setting (the left tank) and departed. The left tank, of course, was the one that contained only about three and a half gallons.

The Safety Board noted that several witnesses to the accident heard an engine backfire or popping noise just before the aircraft plunged towards the water. The checkout pilot informed investigators that when the engine was stopped by shutting off its fuel supply at the end of a flight, loud popping usually occurred. The Safety Board found that each of the above points consistently pointed to the conclusion that the engine lost power because of fuel starvation or exhaustion.

Loss of control

Investigators used another Long-EZ in a ground simulation to see how difficult it would be to reach the fuel selector handle on N555JD. Since the handle was in the proper location on the exemplar aircraft, the investigator simulated reaching for where it had

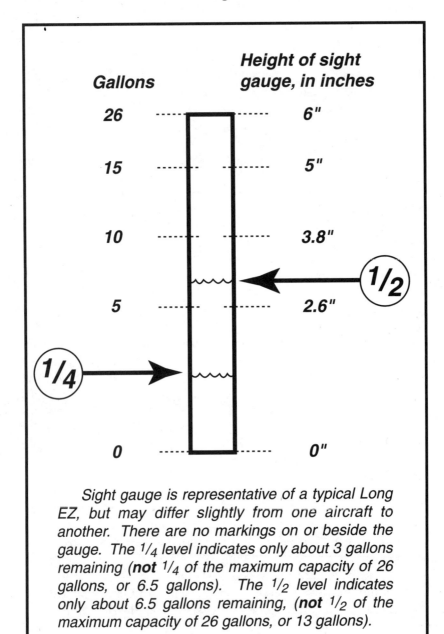

17-4 *Typical Long-EZ fuel sight gauge. . .* Source: NTSB

been located in N555JD. When reaching for the simulated valve, the investigator (who was about John Denver's height) had to completely release the flight control stick, loosen his shoulder harness, turn in his seat while stretching over his left shoulder with his right hand, and then rotate the handle to a nonmarked position. Tellingly, in each simulation the investigator inadvertently pressed on the right rudder pedal while turning to reach the fuel selector handle.

The NTSB noted that a representative from Scaled Composites, Inc. reported that the rudders are very effective on the Long-EZ because of the long moment arm, and that activation of a rudder will produce a pitch-up moment along with the yaw. "On the basis of the accident circumstances and witness accounts, the Board concludes that the accident pilot probably inadvertently applied right rudder while manipulating the fuel selector, precipitating the loss of control that preceded the accident," said the NTSB.

Enhancing safety

The Safety Board expressed concern that the aircraft's builder introduced a safety hazard through his design change of relocating the fuel selector handle. They further commented on the lack of placards for the fuel selector handle and the lack of calibration markings on fuel sight gauges. "Because N555JD was an experimental, amateur-built airplane, specific deviations from the original plans did not require FAA approval, nor was a placard required to identify fuel selector valve positions or to indicate that there had been a change in the airplane's design." They also noted that there are no specific FAA requirement to have inspections for presence of placards and instrument markings for experimental, amateur-built aircraft.

Another concern of the Safety Board was the limited amount of transition training that the pilot received in the Long-EZ. The NTSB revealed that he received about one-half hour of ground school and then about ten minutes of flight training before being allowed to fly solo to Monterrey. The Board stated, "On the basis of his limited flight experience in this type airplane, the Safety Board concludes that the pilot likely did not have the necessary knowledge and skills to efficiently operate it during the emergency circumstances of the accident flight."

The Safety Board pointed out that there are only a few amateur-built airplane training organizations that have published training syllabi. Three companies that sell kits for amateur-built airplanes provide formal, type-specific ground and flight training for their products. The Safety Board also noted that some insurers of amateur-built airplanes, similar to or more advanced than the Long-EZ, have sometimes required formal training as a condition of providing insurance coverage for the pilot/owner.

The probable causes

The NTSB determined that the probable causes of this accident were "the pilot's diversion of attention from the operation of the airplane and his inadvertent application of right rudder that resulted in the loss of airplane control while attempting to manipulate the fuel selector handle. Also, the Board determined that the pilot's inadequate preflight planning and preparation, specifically his failure to refuel the airplane, was causal. The Board determined that the builder's decision to locate the unmarked fuel selector handle in a hard-to-access position, unmarked fuel quantity sight gauges, inadequate transition training by the pilot, and his lack of total experience in this type of airplane were factors in the accident."

Recommendations

In response to the investigation, on February 3, 1999, the NTSB made the following recommendations to the FAA:

- Amend FAA Order 8130.2C to specify that before the issuance of special airworthiness certificates, experimental, amateur-built airplanes should be inspected for needed placards and markings on cockpit instruments and for the appropriate placement and operation of essential system controls to ensure that they provide clear marking, easy access, and ease of operation. (A-99-5)

- Amend FAA Order 8130.2C to specify that the annual "condition inspection" of amateur-built airplanes include an inspection for needed placards and markings on cockpit instruments, and an inspection of the appropriate operation of essential controls to ensure they provide clear markings, easy access, and ease of operation. (A-99-6)

Additionally, they issued a recommendation calling for the FAA, Experimental Aircraft Association (EAA), and Aviation Insurance Association (AIA) to establish a cooperative program that strongly encourages pilots transitioning to unusual or unfamiliar amateur-built, experimental category airplanes to undergo formalized, type-specific transition training (A-99-7, A-99-8, A-99-9).

Industry action

FAA Order 8130.2C, Airworthiness Certification of Aircraft and Related Products, is that document used by the authorities to approve original and recurrent airworthiness certification of amateur-built aircraft. Those airplanes are constructed solely for the pilot/builder's own educational or recreational purposes, and the FAA believes that their role as regulators is only to ensure that "acceptable workmanship methods, techniques, and practices" are used, and to issue "operating limitations necessary to protect persons and property not involved in amateur-built activities." Since there is no regulatory basis, therefore, for FAA approval of placement of cockpit instrumentation, markings or placement or operation of essential system controls on these aircraft, the FAA will not amend the Order.

To meet the intent of the NTSB's recommendation, however, the FAA will revise Advisory Circular (AC) 20-27, Certification and Operation of Amateur-Built Aircraft. This revision will advise builders of "the need to include applicable placards, clearly mark cockpit instrumentation, and consider the appropriate placement and operations of essential control systems...." Furthermore, builders will be reminded of the checklist contained in AC 90-89, Amateur-Built and Ultralight Flight Testing Handbook, which provides appropriate inspection methods for cockpit instrumentation and systems controls.

Epilogue

It is too soon to see how else the FAA, EAA and AIA will respond to the NTSB's recommendations. When it gets right down to it, though, every element of the aviation system must work together to ensure that safety is maintained. It starts with the designer and must carry forward to the builder. The regulatory authorities, associations, and insurers must insist that aircraft are properly designed, built, and maintained. But even the best designed, built, and maintained aircraft are only as good as the pilot who flies them; nothing can overcome the hazards created by a pilot who is not adequately trained or proficient, or who does not tend to proper preflight planning and preparation.

References and additional reading

Castro, Peter, and Bane, Vickie. 1997. Peaks and Valleys. *People.* 27 October, 82-88.

Federal Aviation Administration. 1999. *NTSB Recommendations to FAA and FAA Responses Report.* <http://nasdac.faa.gov>.

National Transportation Safety Board, 1999. Public docket. *Long-EZ, N555JD, Pacific Grove, California, October 12, 1997. LAX98FA008.* Washington, D.C.: NTSB.

North, David, Editor-In-Chief. 1997. Entertainer John Denver. *Aviation Week & Space Technology.* 20 October, 19.

18

The investigation continues...

When reading a book about aircraft accidents, there are certain advantages to having it authored by experienced aircraft accident investigators. After all, each investigation is unique. Each has its own problems, and each its own high and low points. A person who has "been there, seen it, and done it," and understands the subtleties of the behind-the-scenes work can best lead the reader through the sometimes-confusing twists and turns typical to most investigations.

Investigative authorities typically amass reams of information during the course of a mishap investigation. It's all there—the relevant, the extraneous, the interesting stuff, the boring stuff. This book's authors have tried to use their knowledge and experience to "cut to the chase" and provide the reader with interesting, accurate, and succinct information for each of the accidents profiled in the previous chapters.

There is a hitch, though. No person actively involved in an aircraft accident investigation can divulge information while the investigation is ongoing. As investigators actively working with the National Transportation Safety Board (NTSB) and the Transport Safety Board (TSB) of Canada on several major ongoing accident investigations, the authors take very seriously their responsibility to safeguard this sensitive information.

Nevertheless, readers are entitled to basic information about recent accidents. The following information has been gleaned strictly from the public affairs offices of the relevant accident investigation authorities to provide readers with general information concerning five recent, high-profile accidents. As each investigation is concluded, more information should be available.

TWA 800—Off the coast of Long Island, New York. June 17, 1996

On July 17, 1996, about 8:45 P.M., TWA flight 800, N93119, a Boeing 747-100, crashed into the Atlantic Ocean off the coast of Long Island shortly after takeoff from Kennedy International Airport. The airplane was on a regularly scheduled flight to Paris, France.

According to the NTSB, "The initial reports are that witnesses saw an explosion and then debris descending to the ocean. There are no reports of the flightcrew reporting a problem to air traffic control. The airplane was manufactured in November 1971. It had accumulated about 93,303 flight hours and 16,869 cycles. There were 230 fatalities."

Because criminal activity was suspected immediately after the accident, the Federal Bureau of Investigation (FBI) was initially very involved in the investigation. Notably, a highly publicized public hearing was held in Baltimore, Maryland, in December of 1997 to review the progress of research being done in many areas, including fuel flammability and aging aircraft systems.

Swissair 111. Off the coast of Peggy's Cove, Nova Scotia. September 2, 1998

On September 2, 1998, at 9:18 P.M. Atlantic Daylight Savings Time, Swissair Flight 111 (SWR 111), a McDonnell Douglas MD-11 aircraft, departed New York's John F. Kennedy International Airport (JFK), en route to Geneva, Switzerland. On board were 215 passengers and fourteen crewmembers.

According to the TSB of Canada, "Approximately fifty-three minutes after take-off, as the aircraft was cruising at Flight Level 330, [33,000 feet] the crew noticed an unusual smell in the cockpit. Within about three-and-a-half minutes the flight crew noted visible smoke and declared the international urgency signal 'Pan Pan Pan' to Moncton Area Control Centre, advising the Air Traffic Services (ATS) controller of smoke in the cockpit. SWR 111 was cleared to proceed direct to the Halifax airport from its position fifty-eight nautical miles southwest of Halifax, Nova Scotia. While the aircraft was maneuvering in preparation for landing, the crew advised ATS that they had to land immediately and that they were declaring an emergency. Approximately twenty minutes after the crew first noticed the unusual smell, and about seven minutes after the crew's 'emergency' declaration, the aircraft struck the water near Peggy's Cove, Nova Scotia, fatally injuring all two-hundred-and-twenty-nine occupants on board."

The TSB of Canada further stated, "To date, the investigation...has revealed heat damage consistent with a fire in the ceiling area forward and aft of the cockpit bulkhead. Both the Flight Data Recorder (FDR) and the Cockpit Voice Recorder (CVR) stopped recording while the aircraft was at approximately 10,000 feet, about six minutes before impact with the water."

American Airlines 1420. Little Rock, Arkansas. June 1, 1999

On June 1, 1999, American Airlines flight 1420, a McDonnell Douglas MD-82, crashed after landing at Little Rock, Arkansas. "There were thunderstorms and heavy rain in the area at the time of the accident," stated the NTSB. "The airplane departed the end of runway, went down an embankment, and impacted approach light structures. There was a crew of 6 and 139 passengers on board the airplane." Twelve persons lost their lives in that accident, including the flight's captain.

Glider accident, N807BB. Minden, Nevada. July 13, 1999

Generally speaking, gliders, also called "sailplanes," are considered to be quite safe. Fatal accidents are uncommon. The FAA allows a student to solo at the age of fourteen years old, compared to sixteen for power aircraft. But when a glider crash claimed two very experienced and highly respected pilots, the feeling within the soaring community was shock, disbelief, and sadness.

No pilot, regardless of experience or skill, is immune to an accident. Bill Ivans, 79, had been president of the Soaring Society of America (SSA), with over forty years of soaring experience. He was the pilot of numerous record-setting glider flights. His passenger, also a highly experienced glider pilot, was Don Engen, 75. Engen retired as vice admiral from a distinguished and colorful aviation career in the U.S. Navy. In 1982, he was appointed by President Reagan to become a member of the NTSB, leaving that post two years later when he was named as FAA Administrator. Upon leaving government service, he served as President of the Aircraft Owners and Pilots Association's (AOPA) Air Safety Foundation. At the time of his death,

Engen was director of the Smithsonian's National Air and Space Museum, and was on the board of directors of the SSA.

About 12:40 in the afternoon of July 13, 1999, Ivans and Engen departed the Minden, Nevada, airport in Ivan's two-place Schempp-Hirth Nimbus 4DM. Unlike most gliders that must be towed to altitude by power plane, this sleek fiberglass $250,000 German-produced glider had its own engine. The Nimbus 4 had an unusually long wingspan—about eighty-six feet from tip to tip, compared to a more traditional glider's fifty-foot span (see Fig. 18-1).

Another glider pilot was flying about 1,000 feet below the Nimbus and noticed it in a high-speed spiral with about a forty-five degree nose-down attitude. After two full rotations, the rotation reportedly stopped and the flight seemed to stabilize. As it stabilized, however, the witness observed the wings bending upward, the wingtips bowing even higher. The outboard wing tip panels then departed from the glider followed by disintegration of the remainder of the wings. The fuselage dove vertically into the ground.

Another witness stated that he observed the glider in a tight turn, as if climbing in a thermal, when it entered the spiral. The witness stated that the glider appeared to be recovering from a spin.

Another glider pilot stated that right after witnessing the accident from altitude, he encountered "very turbulent" conditions with a sudden forty-knot airspeed increase. He characterized the conditions as "more choppy than normal" and further stated that "controllability was not unmanageable but was rough."

John F. Kennedy, Jr. Off the coast of Martha's Vineyard, Massachusetts. July 16, 1999

If the glider accident involving Don Engen and Bill Ivans got the attention of the soaring community, an accident three days later would captivate the entire world. About 9:41 P.M., a Piper PA-32-R301, Saratoga II, N9253N, was destroyed during a collision with water approximately seven and a half miles southwest of Gay Head, Martha's Vineyard, Massachusetts.

Unique about this accident was that the pilot, John F. Kennedy, Jr., was fatally injured along with his wife, Carolyn Bessett Kennedy, and Carolyn's sister, Lauren Bessett. According to the NTSB, the flight de-

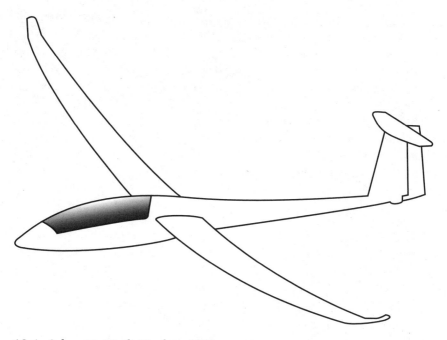

18-1 *Schempp-Hirth Nimbus 4DM...* Art: J. Walters

parted Essex County Airport (CDW), Caldwell, New Jersey, and was bound for the Martha's Vineyard Airport (MVY), Massachusetts. After reaching MVY, the plans were to drop off Lauren Bessett and then continue on to Hyannis, Massachusetts.

Night visual meteorological conditions prevailed, and no flight plan had been filed. The NTSB determined that the pilot obtained an Internet weather briefing earlier that evening, with a route briefing from Teterboro, New Jersey, to Hyannis, with MVY as an alternate. The Hyannis forecast called for winds from 230 degrees at ten knots, visibility six miles, and sky clear, with winds becoming 280 degrees at eight knots. There were no hazardous weather advisories issued for the route of flight, and all airports along the route of flight reported visual meteorological conditions.

The flight departed CDW at 8:38 P.M. The pilot informed the CDW tower controller that he would be proceeding north of the Teterboro Airport, and then eastbound. There is no record of any further communications between the pilot and the air traffic control system. According to radar data, the airplane passed north of the Teterboro Airport, and then continued northeast along the Connecticut coastline at 5,600 feet, before beginning to cross the Rhode Island Sound near Point Judith, Rhode Island. The NTSB's review of the radar data revealed that the airplane began a descent from 5,600 feet about thirty-four miles from MVY. The airspeed was about 160 knots, and the rate of descent was about 700 feet per minute (fpm). About 2,300 feet, the airplane began a turn to the right and climbed back to 2,600 feet. It remained at 2,600 feet for about one minute while tracking on a southeasterly heading. The airplane then started a descent of about 700 fpm and a left turn back to the east. Thirty seconds into the maneuver, the airplane started another turn to the right and entered a rate of descent that exceeded 4,700 fpm. The altitude of the last recorded radar target was 1,100 feet.

The aircraft wreckage was located four days after the accident in 116 feet of water, about a quarter of a mile north of the position where NTSB specialists predicted it would be found. The engine, propeller hub and blades, and entire tail section were recovered. The entire spans of both main wings' spars were also recovered. Additionally, about seventy-five percent of the fuselage/cabin area, eighty percent of the left wing structure, and sixty percent of the right wing structure were recovered. The recovered wreckage was taken from a debris field that measured 120 feet long and was oriented along an approximate bearing of 010/190 degrees.

Preliminary examination of the wreckage revealed no evidence of an in-flight structural failure or fire. The right wing structure exhibited greater deformation than the left wing structure. Two of the three landing gear actuators were recovered and found in the fully retracted position. There was no evidence of conditions found during examinations that would have prevented either the engine or propeller from operating. Visual inspection of the propeller indicated the presence of rotational damage. The NTSB has plans for a detailed examination of the navigation and communication radios, autopilot, and vacuum system (which powers the aircraft's "artificial horizon").

Mr. Kennedy received his private pilot certificate in April 1998. He did not possess an instrument rating. Interviews and training records revealed that the pilot had accumulated about 300 hours of total flight experience. Pilots who had flown over the Long Island Sound that evening were interviewed after the accident, reporting that the in-flight visibility over the water was significantly less than anticipated.

Conclusions

These mysteries will be solved. The investigators working on these accidents are among the best in the world, and their search for the truth will be fruitful. The skies are safe, and through their efforts will remain that way.

References

North, David. Editor-In-Chief. 1999. Engen dies in sailplane crash, headed air and space museum. *Aviation Week & Space Technology.* 19 July, 35.

National Transportation Safety Board. 1999. *Web site home page.* <http://www.ntsb.gov/>

Transportation Safety Board of Canada. 1999. *Web site home page.* <http://bst-tsc.gc.ca/>

Index

About the Authors

James Walters is a Captain with a major U.S. airline and is Chairman of the Accident Investigation Board for the Air Line Pilots Association, International. He is completing studies toward a master's degree in aeronautical science/safety from Embry Riddle Aeronautical University and is a graduate of the Transportation Safety Institute. He frequently participates in National Transportation Safety Board accident investigations and disaster preparedness training exercises conducted by airports worldwide.

Robert Sumwalt is a Captain for a major international airline. Having received training as an aircraft accident investigator from the NTSB, he participates in air carrier accident investigations and also in aviation safety research. A consultant to the NASA Aviation Safety Reporting System, he is the author of numerous aviation articles, training texts, and manuals on aircraft operations, human factors, aircraft accident investigation, and airline policies and procedures.